C0-BVE-453

ENVIRONMENTAL STEWARDSHIP

Images from Popular Culture

Dorothy J. Howell

BERGIN & GARVEY
Westport, Connecticut • London

622846

Library of Congress Cataloging-in-Publication Data

Howell, Dorothy J.
    Environmental stewardship : images from popular culture / Dorothy
J. Howell
        p.   cm.
    Includes bibliographical references and index.
    ISBN 0–89789–391–3 (alk. paper)
        1. Popular culture—United States. 2. Environmental psychology—
United States. I. Title.
NX180.S6H72   1997
304.2—DC21   96–51614

British Library Cataloguing in Publication Data is available.

Copyright © 1997 by Dorothy J. Howell

All rights reserved. No portion of this book may be
reproduced, by any process or technique, without the
express written consent of the publisher.

Library of Congress Catalog Card Number: 96–51614
ISBN: 0–89789–391–3

First published in 1997

Bergin & Garvey, 88 Post Road West, Westport, CT 06881
An imprint of Greenwood Publishing Group, Inc.

Printed in the United States of America

∞™

The paper used in this book complies with the
Permanent Paper Standard issued by the National
Information Standards Organization (Z39.48–1984).

10 9 8 7 6 5 4 3 2 1

Copyright Acknowledgment

The author and publisher gratefully acknowledge permission to reprint the following
material.

Greek Literature in Translation. Whitney Jennings Oates and Charles Theophilus
Murphy, eds. 1944. Reprinted by permission of Addison-Wesley Educational Publishers,
Inc. Excerpts from *The Bacchae.*

For Rick, Bruce, Diana

Contents

I
Origins: Whence We Come

II
Imagination and the Arts: Where We Saw Ourselves

III
Reflections of the Natural World: What We Left Behind

Preface

Much of the text that follows is in the first person plural. Who "we" are, however, depends on context. In terms of prehistory, we are humankind at large. In terms of history, we are living in a society largely dominated by Western European culture. The premise is that most readers, even those not of European extraction, are living in a world strongly colored by the history and traditions of Western civilization carried here by Europeans and still extant in the United States as cultural icons, often appearing in the popular arts.

In a larger sense, we are humankind seeking from each other a personal articulation of what is identified here as environmental stewardship. For some, stewardship is a relationship resident within. For others, stewardship calls for a restoration of an intimate relationship with the natural world. A guiding principle of this work is that the knowledge and insights underlying stewardship should be accessible to any receptive mind. Furthermore, stewardship is ultimately an intensely personal relationship expressed to some extent by all. No one individual, discipline, or culture enjoys a monopoly over this intimate relationship with the cosmos.

The mission here is to open doors onto stewardship for each reader. Hence, the sources are in large part relatively popular or relatively accessible books and articles. If something in the text strikes a chord with a reader, he or she may go to the source and from there proceed to the underlying scholarship, to the subject of that scholarship, or to some corresponding internal image, evoked or instilled.

Introduction

Writers and other artists have long been providing their audiences with images conveying elements of personal identity. Among those images are certain lineages — archetypes — of special places and the "Simple Life." Some of the most evocative images reflect our ambivalence in maintaining a pragmatic balance between our culture-bound persona and an atavistic longing lingering since the emergence of our earliest cities. Today such images are prominent in works of popular culture. Of the popular media, film — cinema and television — reaches immense audiences. With its uniquely intense focal point of sight and sound, theatrical film can be compelling through visceral thrill or subtle charm. In that capacity, film is a metaphor for our relationship with the natural world as that medium adds to the lore of the relationship.

LOCAL HERO OR *ON DEADLY GROUND*

Two films, *Local Hero* and *On Deadly Ground*,[1] share a common theme, which can be broadly characterized as environmental stewardship. The protagonist in each is faced with the plans of a major firm that will devastate a pristine environment. Each man deals in his own way with the plans and the magnate who is the primary mover within the firm. In the end both firms' devastating plans are foiled. Analysis of the two films opens the way to the relevance of both their shared theme and their distinctive treatments of that theme.

Local Hero's MacIntire is a disconnected urbanite, caught up in grid-lock, with his high-rise apartment, frenetic music, and expensive car. He communicates by telephone and through glass, never in person. He even eats from vending machines. His sole connection with the natural world seems to be weather reports, heard on the car radio or through earphones while jetting to Scotland and a remote fishing village on behalf of the "acquisition of Scotland" by Knox Oil. Everything around MacIntire is controlled by technology, including the modeled environment in the laboratory he visits upon his arrival in Scotland: "welcome to our little world." In short, MacIntire is utterly disconnected from the natural world and humanity.

The villagers are a different kind of people altogether, and MacIntire encounters an array of archetypes, including the Goddess and the Trickster. This Knox owns the beach on which he lives the life of a scavenger. Gradually MacIntire reconnects with the animal world, the sea, and the land. The viewer identifies with him as his estrangement from the world gradually wears away, and he makes his reconnections. *Local Hero* is profound but subtle and rich with understated humor. In the end it is melancholic: neither MacIntire nor the urbane viewer is likely to be able to maintain the reconnections that have been established.

This film is rich in irony. The oil magnate, Happer, is utterly bored with the machinations behind his wealth. Indeed, he is already connected, not necessarily with humankind, but with the cosmos, through his fascination with astronomy. On the other hand, Happer's efforts to assure himself of his humanity through his therapist are doomed to failure, indeed the effort is literally as well as metaphorically "perverse." His place is with the Scot Trickster, with whom he is instantly at home on their first meeting. Thus, Happer connects with "simple" people of the Scottish village and acquires the land, whereupon he renounces "MacIntire's plans" for the refinery and oil storage facility in favor of an observatory and marine studies institute.

Throughout the film the wonder of the natural world and a deep need to be a part of it are brought home to MacIntire and the viewer — openly through MacIntire's desire to please Happer by describing the aurora borealis, and more subtly by the insidious impressions conveyed by the villagers, the sea, and the land. Like too many of us, MacIntire is initially oblivious to the natural beauty that surrounds him, but, given time, his connection is restored, at least briefly. The message of the essential need for a personal environmental aesthetic is not only in the evolving plot line but also in every aspect of the film. In short, *Local Hero* evokes something deep within us as individuals. It is as subtle as nature itself can be.

Nature is not always subtle. Sometimes it touches us more with its bludgeons than through nuance. The message of *On Deadly Ground* is overt, even blatant. Its message is one of brutality, violence, and pyrotechnics — for all its purported call for environmental stewardship. Jenning, head of Aegis Oil, is the nemesis of protagonist Forrest Taft. Jenning's mission is to rush an Alaskan refinery into service, no matter the risk to humankind or nature. Taft will single-handedly halt this mission, and he will literally meet fire with fire. Never will he move slowly where he can rush — communication is by computer and travel by helicopter or snowmobile. The viewer has to wonder whether he knows how these machines are powered and whether he has any cognizance of his own disconnection from the natural world and humankind.

The opening scenes of *On Deadly Ground* are promising. The camera pans along scenes of Alaska's wilds. But the viewer is instantly faced with the frantic efforts to bring a screen-filling conflagration under control. Taft, lone hero-savior, descends by helicopter to save the day. This is the archetypal American hero, a man of outstanding stature; strong, even laid back; in remote silence until provoked to violence. That Taft is also intended to be the noble savage, if not an archetypal wild man, is conveyed by his buckskin jackets, Indian blanket, and fur-trimmed clothing.

Although it may be clear that *On Deadly Ground* is intended to reveal a connection among the natural world, primal peoples, and environmental stewardship, neither the progress of Taft's "sacred journey" nor its connection with environmental amenities is particularly lucid. Moreover, any image of the Simple Life and of connectedness to the natural world or to humanity is constantly contradicted.

Taft himself remains disconnected from all around him. He is every bit as estranged from the natural world and humanity as Jennings. Even the attempts to invoke primal images remain vague in intent and disconnected from the action and the viewer. Emerging from primal image and the charge of a shaman, Taft mounts a snowmobile, arms himself with a veritable arsenal, trades the snowmobile for a more powerful truck and thus proceeds on this "sacred journey."

When at last the carnage is done, a body count pointless, Taft emerges unscathed to indulge in a dissertation on environmental stewardship addressed to the fictional press and the real viewer as the latter are treated to horrendous images of oil-soaked birds and mammals set against a litany of statistics chronicling the devastation created by chemicals and oil.

In the end, *On Deadly Ground* is composed of far more fire than earth, air, and water. If the film has one outstanding fault in the context of environmental stewardship, it is that its hero is as ignorant of the harm he is inflicting through his profligate use of technology and power as are most of us. Criticism is really beside the point. It is better to ask which of these two films conveys to an audience an effective message of environmental stewardship.

Of course, they both do. Each communicates to its selected audience, one through nuance and subtlety, the other through violence and bold, startling images of cinematic technology. And communication, in the end, is the principal role of a culture's sciences, crafts, and arts, notably including those of popular culture.

While some of us are moved by nuance, others are reached by bludgeon. And prevailing mood is not to be dismissed: a calm sea framed by gently waving palms and the stupendous fury of a raging storm can be equally evocative.

Throughout all cultures — historical, primal, "advanced" — expressions of the intellect express through or for each individual, directly or indirectly, what is important to us. Whatever we lack in talent for personal expression, we rely on our culture's sciences, crafts, and arts to communicate for us. Recent history and social change may be depriving many of us of mythic or archetypal images, but they are not so easily cast aside. If nothing else they emerge anew, if only in perverted form, in our popular arts.

EXPRESSION

Evolution of the Need

It is the theme of *Environmental Stewardship* that each of us is either endowed with or searching for an environmental aesthetic. If we cannot find a means to articulate that aesthetic for ourselves, we are searching our culture for its expression. Upon experiencing or actually articulating a personal environmental aesthetic, we are better able to commit to a restored connection to the natural world, and, parenthetically, to humankind as well. Ultimately, this is the "sacred journey" that can promote a more effective national environmental policy. Environmental stewardship is how that journey is identified here.

The dictionary defines a steward as a keeper, one in charge of the affairs of a large estate or the supervisor or administrator of the property of another. A biblical dictionary carries that definition further to add a metaphysical implication. In the spiritual realm a steward bears a divine commission and will ultimately be called to an accounting of the steward-ship obligations. The Christian steward is entrusted with certain divine gifts demanding faithful and wise stewardship, arguably including stew-ardship of the earth. Notably, a steward need not stand outside that which is protected but may be a dweller within it.

Many commentators on contemporary American society are reasserting through a variety of expressions our profound relationship to the land. We are reminded to hold the land in regard, to sense "its weather and colors and animals," and to experience its wisdom. We are called upon "to be alert for its openings, for that moment when something sacred reveals itself within the mundane, and you know the land knows you are there."[2] We are reminded of a basic human need for nature and urged to reassert our lost sensitivity to the harvest or the hunt or the natural cycles or simply the *awareness* of the world around us.[3] All the disciplines of human endeavor are engaged in seeking to express some environmental aesthetic, a highly personal articulation of an environmental ethic.[4] Throughout Western history the expression emerges, fades, and emerges again in certain ways: the Simple Life and the sense of place.

Through these expressions we are reminded, often with great sublety and perhaps without any such intent on the part of the purveyor of the image, that we are inseparable from the web of life. In fact, we as a race undergo an "unfolding of consciousness" ranging through the archaic, magical, mythic, mental, and integral.[5] According to this scheme we have progressed from the point at which our consciousness was not yet distin-guishable from the animal, into a unity with all through emotion, to an awakening of the soul, to the point where "will, duality, reconciliation through synthesis, and future-directedness" emerged, to our present "supra-rationality."[6] It is here that we seem to have separated ourselves from the natural world and are suffering the consequences of disconnection and an associated estrangement from the cosmos and from our fellow humans.

Nearly ubiquitous, along with sense of place and versions of the Simple Life, is a deep ambivalence. We want both worlds: the richness of civi-lization and the simplicity of the pastoral, atavistic, or frankly wild. We need to restore the connection. It is not enough that it is being expressed on our behalf through the Simple Life or a sense of place.

Basis for Restoration

Relationships with the natural world are indeed personal, but they arise in a culture, set in a place and time. The human pedigree began some four billion years ago with the origin of life on this planet. It is at once humbling and fascinating to comprehend that we and all of life are indeed one, the descendants of the same initial protoplasm.[7] Indeed, this is the first of numerous occasions when the themes of this book are identified with a web or, especially, a tapestry. In this case it is life that is "an intricately woven tapestry" from which it is never safe to "yank out a few threads here and there" for fear "the whole fabric will . . . unravel." As a result, the themes of mutuality, respect, and interlocking needs are not merely ideology of one faction of society but a biological reality.[8]

Thus, our first culture is a biological one inescapably shared with all of life. But beyond that, we cannot escape our more immediate upbringing and environment, our culture. A principal thread in the tapestry is, however, that we can still search out truths and values beyond our own culture.[9] Every culture, not the least our own, is "a repository of some good thought about the universe"[10] and it behooves us to explore and take to heart the lessons available. Stewardship is an interdisciplinary endeavor of heroic proportions.

Within our culture and through what we can assimilate from others, each of us individually is also a component in that continuum of nature. Our identity includes our adaptation to the natural world at numerous levels, the metaphysical or spiritual as well as the physical or material realm.[11]

We turn to the arts, crafts, and sciences of our culture and beyond to inspire and to inform our personal articulation of connection or reconnection and reconciliation with the natural. The culminating articulation forms our personal environmental aesthetic, which governs our practice of environmental stewardship. In the company of others, that corporate aesthetic infuses our institutions in the ultimate practical application characterized as a national environmental ethic — itself expressed in national programs of pollution control, environmental protection, and natural resources management. Today, much of the mythos of the past and of the arts and literature of a culture are replaced with popular arts for which the two films identified above stand representative.

The Call for Restoration

Today it is increasingly and more forcefully asserted that we must again acknowledge our kinship with nature. Moreover, in doing so we cannot dismiss our emotions and the other "nonintellectual" components of our being. We are a whole as nature is a whole and we are within both.[12] Interestingly enough, it is essential to turn for guidance to the natural sciences such as ecology, to the arts, and to the spiritual realm. And that integrated guidance must be increasingly conscious.[13]

Certainly, the integration of realms is far from new. Rather than something novel, it is something we have lost and now are commencing to regain. Cosmogony (myths, symbols, ritual) and cosmology ("natural history") have informed and interacted with each other from human beginnings.[14]

Personal Articulation of Stewardship

There is no doubt that each of us ultimately works out a personal system of concepts.[15] One of our sources is the collective unconscious. Even maintaining a small garden is part of myth: "A collective phenomenon, myth works, as anthropologists have shown, by reconciling antithetical cultural ideas in a way not always recognized by members of that culture."[16]

Humans have been said to need "liberty, spontaneity, nakedness, mystery, wildness, wilderness." We also need heroes and heroines.[17] Nowhere are those needs more acute than in seeking to articulate an environmental aesthetic. Consider that we may actually be what we "read in the morning." The investor reads the money pages of *The Wall Street Journal.* The commuter listens to the news. The gardener or farmer judges the day in order to make essential decisions. And the primal hunter reads the sky and winds, feels barometric pressure and humidity, observes animals and birds.

How much more our day would be, were we to follow the lead of the hunter in observing the natural world before embarking on our own routine. What a fresh perspective we would carry with us.[18]

THE MISSION

Whatever sources each of us turns to, the mission here is to instill, evoke, or reinforce overt articulation of that environmental aesthetic within each of us. The author's personal articulation comes from being a participant in, and student of, Western civilization as manifested in the contemporary United States. The sources chosen here are her own, as the sources of each reader must be his or her own. The author's sources reflect a combination of vocation in the biological sciences and in evolving environmental law and policy, of avocations ranging from literature to contemporary science fiction and fantasy and fascination with popular film and television, and of an active vocation in Christian theology with a firm foundation in the teachings and predictions of the Old Testament.

While the author's sources are personal ones and limited by her interests, they can inspire each reader to explore his or her own array of sources. The potential resources are literally boundless, and the array identified by each individual almost certainly unique. The author's sources reflect, as will each reader's, her intellectual level as well. But no one is precluded from stewardship; each of us is invited to articulate it personally, wherever we may stand along the intellectual continuum. After all, "[t]he aim of the cosmic questions of both philosopher and child is to obtain . . . some reflexive action from the external world to the self, in order to understand the world in terms of [one's] own experience as well as through cultural explanations."[19]

Terry Tempest Williams urges us to remove our masks and step around our persona to "admit we are lovers, engaged in an erotics of place. Loving the land. Honoring its mysteries. Acknowledging, embracing the spirit of place." She asserts, "That is why we are here. It is why we do what we do. There is nothing intellectual about it. We love the land. It is a primal affair."[20]

NOTES

1. *Local Hero* (1983), Bill Forsyth, director; *On Deadly Ground* (1994), Steven Seagal, director.
2. Lopez, *Arctic Dreams* (1986), 228.
3. Michael Martin McCarthy, foreword to Shepard, *Man in the Landscape* (1967, 1991), xii–xiii, xvi.
4. Ibid., xxiii.
5. Kealey, *Revisioning Environmental Ethics* (1990), 2, 4–5.
6. Ibid., 5.
7. Sagan and Druyan, *Shadows of Forgotten Ancestors* (1992), 131.
8. Ibid., 136, 138–139.
9. See, e.g., Ferkiss, *Nature, Technology, and Society* (1993), viii.
10. Lopez, *Arctic Dreams*, 40.
11. Cobb, *Ecology of Imagination in Childhood* (1977), 100–101.
12. Midgley, *Beast and Man* (1978), 196.
13. Cobb, *Ecology of Imagination in Childhood*, 67.
14. See, e.g., Bron Taylor, "Earth First!'s Religious Radicalism," in Chapple (1994), 185–86.
15. See Midgley, *Beast and Man*, xiv.
16. Machor, *Urban Ideals and the Symbolic Landscape of America* (1987), 16.

17. Williams, *An Unspoken Hunger* (1994), 75. The author is quoting Edward Abbey during a conversation with him in Utah.

18. Meeker, *Minding the Earth* (1988), 2–3.

19. Cobb, *Ecology of Imagination in Childhood*, 28.

20. Williams, *An Unspoken Hunger*, 84.

I

ORIGINS:
WHENCE WE COME

1

Origins of the Western Tradition

THE NATURE OF CULTURE

We are the culmination of our shared past, Western Civilization. This is an essential point of departure for exploring contemporary environmental stewardship, because our way of life as individuals and as a nation is bound to our stories, our choices, and even the directions and numbers we prefer. And these elements together reflect not only our view of the natural world but also the array of assumptions we make about the world around us.[1] Even if we choose to explore beyond our tradition in establishing our environmental aesthetics, we should first examine what our tradition has to offer. To set the stage, it is appropriate to establish working definitions.

Nature can be defined in terms ranging from the biological through the instinctive, raw, or unconscious to the primordial. In contrast, culture can be defined in countervailing terms ranging from order through consciousness to deliberation and development. In this context aesthetics emerge from roots suggesting "shaped feeling" and "sensitive perception."[2] In less dramatic terms, nature is the material realm about us, culture is the human setting in which the individual most immediately interacts with other humans and with the world, and aesthetics represent an appreciation for one's surroundings.

We humans are inescapably of and in nature: We are biological entities, but we are incomplete without the "programming" effected by our culture.

That culture is one way in which our faculties are awakened.[3] In biological terms, culture represents a novel level of organization founded in the mind.[4] Humans are at the center of a biological continuum starting with the subcellular and proceeding in stages from cells through tissues, organs, organ systems to the organism itself and on to the population of like organisms and ultimately to the community of all interacting organisms.

Culture is what renders humankind more than some featherless biped. Culture is what leads us to "smell the sea, listen to the silence in the intervals of music, contemplate the shifting spaces of an achitectural interior, marvel at what can be seen only with the mind's eye — the curvature of the universe." Unfortunately, culture has also been identified as the human attempt to rise above nature.[5] Thus is presaged the profound ambivalence between nature and culture.

Culture further provides us with symbol, ritual, and ceremony by which we begin to express our relationship to our setting. A significant part of that setting goes beyond the material realm into the metaphysical where we find ethics, philosophy, and religion as well as other forms of the *super*natural.[6] The aesthetics evoked by nature filtered through our culture make us more or less attuned to the beauties of the world. Without aesthetics, we lack an essential, to the extreme of enduring a "living death."[7]

Throughout history and throughout cultures humankind are seeking to balance kinship with nature against mastery over nature. The heroic imagery of conquest hardly commences with Saint George and the dragon but delves deep into our past in Babylon, Egypt, and the Greek city-states.[8] No culture, however urbane or primal, has been or is wholly benign in its inevitable shaping of its ecosystem. Devastation has been wreaked at the hands of even primal cultures, since ancient times and on a global scale.[9] This realization lends perspective to the modern condition and may alleviate some of the paralyzing guilt we endure.

If environmental devastation is not new, neither is its counterpart, environmental stewardship. This concept dates as far back as Rome's philosophers, following the Greeks' Plato but before the prevalence of the Judeo-Christian tradition. Plato in *Phaedrus* asserts, "It is everywhere the responsibility of the animate to look after the inanimate," and the Judeo-Christian tradition holds that "Man is to nature . . . as God is to man"[10] — in other words a guardian or steward.

Still, we persist in self-criticism, a healthy exercise; from the times of Socrates and Christ through those of Rousseau and Marx, we of Western culture have been so engaged. Indeed, we are "not a single culture at all, but a debating ground, not a monolith but a fertile confusing jungle of sources."[11]

At this juncture, it is crucial to acknowledge that we all have an active role. In any culture only a few are "at ease verbalizing the group's values." They are often "the shaman, the medicine man, the courtier, or the scholar." Nevertheless, *all members of a group who participate in its ceremonies may be presumed to have a feeling for what is going on.*"[12] These observers are our surrogates. We as individuals can learn from them, but they are not the sole possessors of either wisdom or aesthetics.

THE MATERIAL REALM:
PHILOSOPHY AND NATURAL HISTORY

Significantly, the classical philosophers and the naturalists were one and the same. They have been described as men with a mission, whose mixture of scientific and metaphysical studies constituted an inspirational "vision of the glory of the universe."[13] Among them, Plato is described as more than a student of ethics and politics; he was also a mystic.[14]

Even early thinkers were perceiving a separation of humans or their culture from nature. Natural phenomena or entities were deemed "more basic, more ancient, and therefore more vulnerable and superior to what is newer and/or made by humans." And while these thinkers discovered nature to be a cosmos governed by inherent laws, they occasionally identified nature as Gaia, an "intellectual organism."[15] It would seem not much has altered over the centuries between these ancestors and the contemporary us.

Prior to Socrates, Greek philosophers were pursuing such problems as the nature of matter, the order and arrangement of the universe, and existence itself. With Socrates, enquiries turned to humankind and inward. It was Plato who described all beings as composed of a charioteer driving a team of winged horses. The team of the human being was made up of one noble-bred steed and one of ignoble breeding. It is no wonder that the driving of them is troublesome at best.[16] This is an early recognition of the profound ambivalence in balancing between nature and culture.

This ambivalence remains a profound influence today, particularly among those who dwell in cities, as was recognized by Aristotle, who said, not that we are a political animal, but that we are a *polis*-dwelling animal. As such, we are neither wholly urban nor wholly rural; both perspectives are essential. We need self-sufficient communities of precisely the right size to balance that self-sufficiency with independence and with the opportunity for full human development.[17]

To Plato there was a distant heaven "of which no poet can sing worthily." In this distant place resides an "intangible essence, visible only to mind," but yet "divine intelligence" is "nurtured upon mind . . . and the intelligence of every soul . . . capable of receiving the food proper to it." Each human soul once beheld that true being but none can easily recall that other world because the "memory of the holy thing which once" was seen has been lost.[18] Plato also spoke of *anima mundi,* the soul of the world, what we now call the biosphere.[19]

Aristotle echoes Plato in the observation that all humankind "by nature desire to know" as is indicated by our delight in our senses going beyond their mere usefulness. To Aristotle wisdom is the knowledge of causes and principles, and the wise person knows all it is individually possible to know. Much later Epictetus exhorted us to behave in life as at a banquet: Politely take the dish that comes to you and do not demand those not offered. And, if we decline to take all that comes to us, we will "share the gods' banquet" and their rule.[20] Certainly, this is a lesson in the moderation for which the classical era is admired.

THE METAPHYSICAL REALM

Neither a glance at Western tradition nor a study of stewardship can be complete without exploration of the metaphysical realm. Any religion in harmony with its time's "creative spiritual energies" as expressed in its myths, rituals, and symbols is powerful indeed. Such a religion touches the heart and awakens faith. And ultimately its "ideas of God and teachings about the relation between God and the world shape human attitudes toward nature."[21] Early religious interpretations of the natural world reveal a deep need to have a part in nature's establishment of that world. This need is expressed in the dramatization of human intuitions about our relationship to nature. The historical evolution of this relationship occurs in autobiography on the personal level and through natural history, biological evolution, and cultural evolution on the larger human scale. All can be characterized as cosmic events.[22]

Classical Metaphysics

It appears that religion in the Roman Empire neither challenged nor provided an answer to the meaning of life. Enlightenment was to be found among the philosophers, not the theologians. Religion of the time was more cult and ritual than ideas or theories. Yet fragile civilization was deemed vulnerable to humanity's god-patrons, who must be pampered and appeased.[23] Into this world had come Judaism and then Christianity, which were followed by Islam.

Judaism and Christianity

The God of Abraham, Moses' I Am That I Am (YHWH, Yahweh, Jehovah), has been described in recent time as rendering humankind a cocreator in a perpetual covenant with Him. Thus, from the era of the covenant with Abraham, Jews confirmed their own creative powers through a kinship with the Creator in a voluntary relationship it is fair to call stewardship.

The Jews were not alone in believing God close to the natural world. Hinduism is seen by some students as a "monotheism with all the Hindu gods being aspects of a single universal entity (the only reality), while the world itself is unreal,"[24] and the Jewish tradition merged with that of the Greek philosophers who studied the cosmos as what we now call naturalists.[25] Philo of Alexandria was Jewish and a Platonist philosopher. To him there was only one way to begin to grasp an otherwise wholly incomprehensible God, and that was through the revelation of His powers (or energies) in the world and the order of the cosmos. These powers to Philo "are the highest realities that the human mind can grasp."[26]

It was the Jews who believed God would become accepted by all peoples, and over some three thousand years the "tradition remain[s] that God has not ceased His creative activity when the world had been made."[27] It is also Judeo-Christian tradition that the original harmony created in the cosmos was broken by humans, who in effect estranged themselves and the world with them from God. The same tradition holds that we are all moving toward "the peace of a reconstituted paradise."[28]

According to rabbis, God is "intimately present within mankind and the smallest details of life."[29] His presence is almost tangible and uniquely experienced by each individual, because God adapts Himself to each according to individual capacity to comprehend. And yet that divine essence remains "shrouded in impenetrable mystery," something yet to be attained. A contemporary Jewish philosopher refers to Judaism as a spiritual process by which believers strive for "elemental unity[,] an encounter with a personal God."[30]

Origen, an early Christian theologian, initially believed that Christians must turn against the world, but his later theology stressed God's continuity with the world.[31] Jews and their Christian counterparts are required to keep their covenants with the Lord. Upon human failure to do so, the land would strike as God's agent. Wanton destroyers of nature are subject to divine punishment.[32]

Islam

Al-Lah is described by Karen Armstrong as more impersonal than YHWH, discernable only vaguely through nature and so transcendent as to be discovered only through parables. The Qu'ran urges Muslims to view the world itself as an epiphany through which the imagination must carry the believer beyond the temporal world to "the transcendent reality that infuses all things." Rational curiosity about the natural world itself is encouraged because its workings can reveal its "transcendent dimension and source." Accordingly, early Islamic tradition holds that al-Lah revealed to Muhammad that He created the world in order to be known, and the Qu'ran "teaches that [al-Lah] had sent messengers to every people on the face of the earth." Indeed, the "self of God" completely surrounds the believer, al-Lah is immanent.[33]

Relevant among the significant teachings of Islamic philosophers during the first millenium is that Gabriel is the Holy Spirit of Knowledge and the source of light as well as knowledge. The human soul holds a practical intellect that relates to the natural world and a contemplative intellect in intimacy with Gabriel.[34] Humankind straddle the physical and spiritual realms of reality, the latter through the act of al-Lah. Our reasoning goes beyond the objective to transcend time and space and partake "of the same reality as the spiritual world."[35]

CULTURAL RELATIONSHIP
TO THE NATURAL WORLD

Our cultural ancestors, the Greeks, intuitively grasped the relationship between body and cosmos and perceived the achievement of a harmonious relationship as the "highest form of social philosophy." In essence the Greek imagination was consciously ecological. It is they who provide our contemporary ecological terminology.[36]

From the foregoing, environmental stewardship appears to be a tenet of the three major monotheistic religions practiced today. Consider the observations of Plotinus less than three hundred years into Christianity: We sense something amiss, we are at odds with self, inner nature, and others.

Conflict and a lack of simplicity seem to characterize our existence. Yet we are constantly seeking to unite the multiplicity of phenomena and reduce them to some ordered whole. . . . This drive for unity is fundamental to the way our minds work and must . . . also reflect the essence of things in general. To find the underlying truth of reality, the soul . . . will have to look beyond the cosmos, beyond the sensible world and even beyond the limitations of the intellect to see into the heart of reality. This will not be an ascent to a reality outside ourselves, however, but a descent into the deepest recesses of the mind.[37]

NOTES

1. See, e.g., Williamson, *Living the Sky* (1984), 299.
2. All terms as defined by Tuan, *Passing Strange and Wonderful* (1993), 1, 8.
3. Midgley, *Beast and Man* (1978), 286, 291.
4. Roszak, *The Voice of the Earth* (1992), 200.
5. Tuan, *Passing Strange*, 240.
6. Ibid., 182–183.
7. Ibid., 1.
8. Sheldrake, *The Rebirth of Nature* (1991), 37.
9. Zeveloff et al., *Wilderness Tapestry* (1992), 188–194.
10. Passmore, *Man's Responsibility for Nature* (1974), 28, 31.
11. Midgley, *Beast and Man*, 295.
12. Tuan, *Passing Strange*, 172.
13. Armstrong, *A History of God* (1992), 92.
14. Ibid.
15. Ferkiss, *Nature, Technology, and Society* (1993), 4.
16. Oates and Murphy, *Greek Literature in Translation* (1944) 438, 500.
17. Boorstin, *The Creators* (1992), 92.
18. Oates and Murphy, *Greek Literature*, 501, 503.
19. Berry, *The Dream of the Earth* (1988), 22.
20. Aristotle, *Metaphysics* (Book I), in Oates and Murphy, *Greek Literature*, 603, 605, 685.
21. Rockefeller and Elder, *Spirit and Nature* (1992), 3.
22. Cobb, *The Ecology of Imagination in Childhood* (1977), 51, 52.
23. Armstrong, *A History of God*, 91–92.
24. Boorstin, *The Creators*, 41–42, 45.
25. Berry, *The Dream of the Earth*, 124.
26. Armstrong, *A History of God*, 69.
27. Boorstin, *The Creators*, 45; Berry, *The Dream of the Earth*, 124.
28. Berry, *The Dream of the Earth*, 124.
29. Armstrong, *A History of God*, 73. See also Ferkiss, *Nature, Technology, and Society*, 12.
30. Armstrong, *A History of God*, 69, 73–74, 386.

31. Ibid., 99.

32. Ferkiss, *Nature, Technology, and Society*, 12.

33. Armstrong, *A History of God*, 143, 150, 152, 160.

34. Abu Ali ibn Sina (Avicenna), in Armstrong, *A History of God*, 181, 184.

35. Abu Hamid al-Ghazzali (Al-Ghazzali), in Armstrong, *A History of God*, 186, 189.

36. Cobb, *The Ecology of Imagination*, 66.

37. Armstrong, *A History of God*, 101–102.

2

Europe in the Continuum of Western Tradition

THE NATURE OF HISTORY

The pun, significantly a common usage, reflects a profound commentary on the chronology behind today's call for a restoration of relations with the natural world. To start at the beginning, primal stories are the earliest versions of world history.[1] These pre-literate ancestral traditions predate the formal discipline known as history.

A notable characteristic of primal stories is the prominent place afforded natural beings from plants and animals to rivers, mountains, winds, and stars. In primal (hi)stories nature plays an active and prominent role in human identity, which, in turn, is placed in overt relationships with "other species and powers." Here humans are not the sole determiners of our own destinies: we are not alone, and the connection with the world is an intimate one.[2]

By way of contrast, despite the lineage from primal story to history, the latter conventionally "relegates fellow beings to the periphery" at best. One has to wonder with J. Donald Hughes whether such an isolated vantage on events is not "just another story." He notes that history is distinguished from fiction through the historian's choice of elements and interpretations. The legitimate historian cannot contravene known facts and must take at least his or her own community into account. Legitimate history *must also make ecological sense!*[3]

The historical lineage of Western tradition has assumed a kind of ascending human development: the "nasty, brutish, short" life of the Paleolithic hunter, the agricultural revolution with its more dependable food supply, the urban revolution and origin of civilization with the progressive Ages of Stone, Iron, Bronze, and Steel. The inference of the inspiration of economic growth and technological improvement is unavoidable. What is ignored in this traditional lineage is the natural world. What is neglected is the matter of sustainablity, notwithstanding the evidence of any number of "false starts" attributed by Hughes to depletion of local natural resources.[4]

Hughes, then, calls for an ecological theme in history. After all, human events are in the context of natural phenomena. It is perhaps too obvious to require statement, but from the outset "[e]conomics, trade, and world politics are regulated by the availability, location, and finite nature" of natural resources. What is less obvious but equally influential is that our "actions can divert, but not retrieve, time's arrow." Historians and cultures alike ignore these realities at their peril.[5]

While generally ignoring the natural world, historians for the most part look at events in the panoramic. Persons are not as important as the events carrying them along. Nevertheless, considering individuals is neither novel nor unique. Edward Gibbon, whose first volume of the *Decline and Fall of the Roman Empire* appeared in 1776, was entranced more by history's people than by its politics. Personal experience led this historian to turn away "from the medieval demand for meaning" — a philosophy of history — to individual roles. As Daniel J. Boorstin describes Gibbon's times, there was a shift to "tiny increments of knowledge," and it became "possible for every man to become his own scientist, and perhaps also his own historian."[6]

In the end, according to Boorstin, "[h]uman habits, utterances, exclamations, and emotions are not mere raw materials for distilling 'forces' and 'movement' but the very essence of history."[7] And so, here the individual, ordinary as well as the somehow extraordinary, is every bit as important as the natural world in the emergence of a personal environmental ethic. A chronology sets the stage for the respective roles of environmental features and of individuals.

CHRONOLOGY

A cursory chronology of European history sets the stage for the historical and political contexts in which the trends particularly important to environmental stewardship emerged. The Renaissance is our point of departure. Ideas as well as events are salient features of this chronology.[8] A pattern of "inevitable sameness" of takings, exploitation, and exhaustion of land is not long in emerging.[9]

Marco Polo embarked upon his remarkable journey to the mysterious East near the end of the thirteenth century. About thirty years later, Dante commenced his *Divine Comedy.* More than two centuries after Marco Polo departed for the East, Columbus reached the West Indies of the equally mysterious West. About two centuries after Dante began his work, the Portuguese laid claim to St. Helena. In the next decade Thomas More published *Utopia,* and three years later Magellen embarked upon his westward voyages. In half a century deforestation had commenced in the West Indies.

Over the seventeenth century the Dutch claimed Mauritius, the British and Dutch East India companies were founded, the Jamestown colony was established, the British settled Barbados, and the Dutch claimed or took St. Helena and Tobago. In the year of that last event the aurochs, a European bison, became extinct in Poland. European claims to the lands to their east, south, and west continued through the century, and in the latter half the French founded their East India Company while the Dutch mandated forest protection in South Africa and the dodo bird became extinct in Mauritius.

Then Came the Enlightenment

In the second decade of the eighteenth century the redwood in St. Helena was afforded protection; European claims, takings, and colonialism continued apace. In 1763 a Treaty of Paris provided the British with St. Vincent. In 1764 forest reserves were established on Tobago, and in 1769 the French passed conservation laws effective in Mauritius.

The year 1776 marked the commencement of the American Revolution, 1780 the onset of years of famines in India, 1789 the French Revolution. In 1791 the Kings Hill Forest Act was passed in St. Vincent. With the advent of the nineteenth century, the British repossessed Tobago and Mauritius.

At the close of the century the French Revolution was followed by war between Britain and France. During both events game in French forests were slaughtered and domestic animals allowed to run free. Destruction of grass and saplings resulted. Because both countries lost ships, thousands of oaks were required to replace them. Forests were denuded; some have never been reforested.[10]

The nineteenth century also brought the age of Romanticism.

Modern science emerged about mid-century. Charles Lyell published *Principles of Geology* in the early thirties, and Charles Darwin landed in the Galapagos in 1835. From then until the end of the decade India suffered another drought. During the fifties British scientists reported deforestation in India, the Madras Forest Department was founded, the Forest and Herbage Preservation Act was passed in South Africa, Darwin published *Origin of Species*, and bird protection laws were enacted in Tasmania.

Through the sixties droughts continued to plague India and afflicted both India and South Africa through the seventies. During the sixties the Indian Forest Service was established and bird protection laws enacted in Britain.

In synopsis, one might fairly call European nations schizophrenic with regard to their cultural relations to the land. By the closing years of the eighteenth century, the colonies in the Americas were entering their own schizophrenia in relation to the natural world. European trends set the stage for their American counterparts, in contemporary terms as well as historical.

ANTHROPOLOGICAL AND POLITICAL CONTEXT

Simon Schama asserts that *environmental* history is "some of the most original and challenging history now being written." From an inevitable sameness of historical takings, exploitation, and exhaustion of land, a penitential mood has arisen. More significantly here, the nature myths Western culture is believed to have sloughed off have never gone away. Contemporary historians have moved from a distanced approach to one more directly experiencing "a sense of place" and utilizing "the archive of the feet."[11]

A central human tragedy of the fourteenth century is the Black Death. Out of this dreadful plague two directions emerged and have lingered into our own time and place. One is metaphysical, the other scientific. Religious redemption from a tragic world is sought by believers. Greater control of the natural world is sought through science in order for humankind to escape pain and to convert elements of the environment into natural resources.[12]

To the Black Death is attributed, if not some uniquely excessive anthropocentrism among Westerners in the twentieth century, then a theocentrism. Preceding germ theory, a loss of a third of the human population was attributed to an untoward attachment to the earth and a concomitant need for "spiritual detachment and absorption into the divine."[13] At this juncture, the relevance of the city cannot be overstated. In fact, the city can represent a "telling symptom" of our "collective state of soul." With not little credibility Theodore Roszak asserts that the city has always "balanced on the edge of psychopathology." As an institution of civilization, the city is dated to the days of the pharaohs and "conquering god-kings." This and Roszak's other descriptions are pointedly relevant to the trends to be explored here: "born of delusions of grandeur," "built by disciplined violence," "*dedicated to the ruthless regimentation of man and nature*"[14] (emphasis added). In fact, Roszak frankly describes the city as pathological, a cultural "body armor" deflecting "close contact with the natural continuum from which we evolve."[15] And yet the city is the setting from which a Louis XIV could be inspired to proclaim in the seventeenth century a forest ordinance that for the first time set about the systematic protection of a nation's natural resources.[16] Thus, the profound ambivalence of so-called civilized peoples is manifested.

MONOTHEISM AND THE NATURAL WORLD

While ambiguities abounded in historical events, monotheism was taking its own path and commencing the divergence that has led to the modern natural sciences. It is ironic that while primeval forests of Europe were disappearing, Gothic cathedrals were being erected on a forest motif. Even so early, the architectural metaphor suggests both nostalgia and profound emotional response associated with the forest.[17]

In the Christian world, Augustine's definition of sin — living a lie — is fundamental to our theology and to our relationship to the natural world as well. Living a lie refers to false relations, not exclusively to God, but in general. The goal in life, according to Augustine, is to take steps "toward health and reality" and toward "living in proper, appropriate relations with all other beings." Augustine's next step is to assume a sense of responsibility, "the universal vocation for all people."[18]

Today, the World Council of Churches calls for the integrity of creation in addition to the more predictable call for peace and justice. Earth, in short, is part of one's vocation, not merely an avocation, because, in part, God is a *continuing* creator.[19] Thus, it seems the voices coming out of deep ecology are actually echoes of the traditional view.

The two major theological traditions were undergoing their own histories concurrent with the political and anthropological history of Europe and Western civilization. These theologies are relevant to stewardship in a variety of ways. In some ways deeply connected to each other, Islam and Judeo-Christian theology speak to the individual's spiritual relationships with the natural world and are the foundation of contemporary natural sciences, through both of which the individual is informed about the need and basis for environmental stewardship.

Islamic Tradition

The Islamic world of the ninth century has been compared to a combination of the European Renaissance and Enlightenment. Astronomy, medicine, and mathematics flourished in the company of alchemy. Muslim philosophers were dedicated to a rationality according to the laws of the cosmos they "discerned at every level of reality." To at least one Islamic sect rationalism was not merely religion, it was religion's most advanced form, an evolution in the notion of God revealed beyond scripture.[20] Muslim philosophers warned against both relegating ah-Lah to a separate intellectual compartment and isolating faith from other endeavors. To such thinkers, science as well as religion could only be a valid path to God. There was no contradiction among revelation, faith, rationalism, and science.[21] Presumably, such beliefs would help avoid the disconnection and estrangement with the natural world that has characterized Western traditions.

Judeo-Christian Tradition

The Judeo-Christian tradition was entering an era when the eternal chain of cosmic events from God, the Creator, through all its links to every human action led to the concept of a fixed cause-and-effect pattern leading to determinism.[22] And yet ninth-century Celtic Christian philosophers perceived God as the essence of all being, every creature a theophany, an indication of God's presence. Among them one observed that the perceivable aspect of God is that "which animates the created world" and is revealed "in flowers, birds, trees and other human beings."[23] Thus, there is ambivalence in this aspect of Western tradition as well as elsewhere.

Different lines of thought emerged among Christians. Francis Bacon held excessive knowledge to be a human downfall and believed that the learned become heretics or even atheists. Francis of Assisi "distrusted books as sources of worldly pride." In contrast to Bacon's determinism and scientific certainty, Francis was a forerunner to romanticism in his "sense of an intimate union with all natural things."[24]

To Baruch Spinosa, the seventeenth century Jewish philosopher, God was defined as the law that orders the existence of all things material as well as spiritual. As the highest perfection, God constituted the "aggregate of all laws in existence" and welded everything into "unity and harmony."[25]

Western Science

From its beginnings in religion, Western natural science evolved the Cartesian view of the world, now including the whole of the observable universe. According to this view, emerging from the late sixteenth into the mid-seventeenth century, all material entities, including biota, are mere objects "rigidly subject to the laws of mathematics and physics." According to Francis Bacon, knowledge of natural phenomena was indeed power, because "[o]nce man knew the secrets of nature, he could bend nature to his will."[26]

By mid-eighteenth century, the natural sciences and associated technologies perceived an "objective world," and any "mystical vision of divine reality" and the "evocation of natural and spiritual forces" were replaced by "scientific attitudes" and "technological inventions."[27]

But with the ascendency of this version of science came a more promising concept, that of "emergence." Every time the mind crosses a boundary in the hierarchy of the natural sciences — say from mathematics to physics or physics to chemistry or chemistry to biology or among the levels of biological systems — authentic novelty is encountered: "Possibilities arise that were not there before and could not have been predicted."[28]

It is this "emergence," a reality of nature, that permits specialization within the sciences without a total loss of interconnections and relevance. Surprisingly, it seemingly restores humankind to at least one kind of centrality in the universe. In qualitative terms life is special. It is a unique kind of complexity, and the universe is "in the business of elaborating complexity."[29] Emergence is a scientific principle representing a substantial material basis for our reconnections with the natural world and even with the spiritual realm, as explored in "new stories."

Historian Lynn White, Jr., (1967) has asserted that among Christian personages, Francis of Assisi should be the patron saint of ecology. Indeed, Francis was so proclaimed in 1980 by Pope John Paul II. To Francis all animals are our brothers and sisters and the kinship went on to include Brother Sun and Sister Moon.[30]

The contemporary version of science came into being in the mid-nineteenth century, arising from the natural philosophy of the preceding two centuries.[31] With time, scientists thrived in an awareness that matter was no longer opaque. They also came to discover that natural phenomena are not extrinsic to humans. In a way the human can be described as the self-consciousness of the evolving universe. But simultaneously, the very complexity of data that can unite humankind with the whole of its environment has further estranged believers and those who are purely secular.[32] That complexity has served also to take science beyond the reach of most individuals.

Both religion and science are trivialized as they are drawn apart. Children are the losers as they discover the earth and the nature of life without any sense of wonder instilled by some form of religious belief. The loss is deepened if one acknowledges schooling as the formal contemporary Western substitute for primal initiation preparation and ceremonies.[33]

One result of the Western separation of theology and science is summed up in the attitude toward humans and the universe attributed to the British. It is an attitude tending toward the "firmly empirical and positivistic: only those things which can be measured may be said to exist, and all other comments on the nature of things — like monotheism and metaphysics generally — tend to be dismissed as mere emotional noises."[34]

But if one reconnects theology and science and sees God as Sallie McFague's "continuing creator," the profound union and interdependence with all other beings change everything. Then the world is perceived from neither an anthropocentric nor a utilitarian perspective or even in dualistic hierarchies. Instead we have "a sense of belonging to the earth, having a place in it along with all other creatures, and loving it."[35]

THE PLACE OF THE INDIVIDUAL

Philosophers, theologians, and scholars may ponder and pontificate about human relationships to the realms of the world around us, but all of us live in the world as we individually perceive and relate to it. While we may not all be scholars, we *are* each our own philosopher and theologian, articulate or inchoate.

In historical context, it has been argued that our "mystical bonding with the natural world" has progressively weakened ever since the devastation of the Black Death and its aftermath of separation between metaphysics and Western science. We "lost our capacity to hear the voices of the natural world . . . [and] no longer heard the voices of" the environments about us. Lost, too, was communication with animals that goes beyond emotion and aesthetics. These "languages," lost to our culture and to us as individuals, "are transformations of the soul at its deepest level."[36]

England, specifically, is a valuable point of departure in that our legal system, much of our national tradition, and many personalities have evolved from those Anglo-Saxon roots. Historical use of wood provides a good example of the tensions at the level of the individual. Because this resource was subjected to consumption for fires, the prospect of its exhaustion caused apprehension, yet legislation as early as the sixteenth century proved ineffectual. The rhetoric of observers could have arisen in our own time in the dismay over the "serious evils from the wasteful economy." The dismay, however, was with the people themselves, not their representatives in government or the marketplace. As George Perkins Marsh noted, few "would tolerate any interference by the law-making power with what they regard as the most sacred of civil rights — the right . . . of every man to do what he will with his own."[37] The ambiguity toward nature and natural resources is not exclusively the province of a culture; it suffuses the individual.

And yet, it is the "web of relationships throughout the universe" of which we are first conscious. And there we find "a supreme mode of communion" within us and "with the human community" and "the earth-human complex." Our "identity is inseparable from this capacity for mutual presence." Our intergenerational mission is "to develop this capacity for mutual presence on new and more comprehensive levels."[38]

NOTES

1. Hughes, "Ecology and Development as Narrative Themes of World History" (1995), 1.
2. Ibid., 1, 2.
3. Ibid., 2.
4. Ibid., 3–4, 5–7.
5. Ibid., 9, 10.
6. Boorstin, *The Creators* (1992), 334, 340.
7. Ibid., 341.
8. This chronology is adapted from Grove, "Origins of Western Environmentalism" (1992), 44–45.
9. See Schama, *Landscape and Memory* (1995), 13–14.
10. Ibid., 180.
11. Ibid., 13, 14, 24.
12. Berry, *The Dream of the Earth* (1988), 128.
13. Swimme and Berry, *The Universe Story* (1992), 199.

14.   Roszak, *The Voice of the Earth* (1992), 219.

15.   Ibid., 220.

16.   Ferkiss, *Nature, Technology, and Society* (1993), 55.

17.   Shepard, *Man in the Landscape* (1967), 172.

18.   McFague, "A Square in the Quilt" in Rockefeller and Elder (1992), 39, 43.

19.   Ibid., 44, 47, 53.

20.   Armstrong, *A History of God* (1993), 170. The sect was the Faylasufs.

21.   Ibid., 171, 173.

22.   Jamil, *Transcendentalism in English Romantic Poetry* (1989), 2–21.

23.   Armstrong, *A History of God*, 199, with reference to Duns Scotus Erigena (810–877).

24.   Fairchild, *The Noble Savage* (1928), 8.

25.   Armstrong, *A History of God*, 311, 312.

26.   Ferkiss, *Nature, Technology, and Society*, 34–35.

27.   Berry, *The Dream of the Earth*, 40.

28.   Roszak, *The Voice of the Earth*, 172.

29.   Ibid., 185.

30.   Ferkiss, *Nature, Technology, and Society*, 21.

31.   Aulie, "Al-Ghazali Against Aristotle: An Unforeseen Overture to Science in Eleventh-Century Baghdad" (1994), 28, 40, note 4.

32.   Berry, *The Dream of the Earth*, 128.

33.   Ibid., 130–131.

34.   Jamil, *Transcendentalism*, ix.

35.   McFague, "A Square in the Quilt," 54.

36.   Swimme and Berry, *The Universe Story*, 199.

37.   Marsh, *Man and Nature* (1864, 1965), 192, 201.

38.   Berry, *The Dream of the Earth*, 135.

3

Tradition and the Founding of a Nation

Simon Schama reminds us that the frontiers between the wild and the cultured — like human past and present — are not easily fixed, and we are the heirs of both our history and our historical relationships with nature.[1] Two ways in which we relate to nature are through landscapes, especially in visual renditions, and through natural history — "anecdotal" observations, visual or literary, of plants and animals. On the practical level, our relationship to the land takes the form of ownership or possession, derived from Roman and feudal concepts. This aspect of land has come to represent civilized concepts of social status, political prestige, and power subject in turn to law, custom, and the tolerance of neighbors.[2]

But there is more to land than possession, and our emotional or aesthetic response to land has its own history, one associated with the resident plants and animals. Along the recent historical path, we see a splitting of approaches into the aesthetic, the scientific, and the spiritual. This disconnection is one symptom of our cultural estrangement from the natural world, which can lead to ambivalence about our relationship to both civilization and nature.

HISTORICAL AND POLITICAL CONTEXT

First Impressions

A pastoral myth is said to haunt the American continent. That myth arose out of the European traditions the pioneers brought with them. To them this continent offered the ideal of living "in simple harmony with abundant nature." Significantly, this ideal is in contrast with what is identified as the "ruthless conquest" of this continent of abundance.[3]

Upon arrival, Europeans discovered an Eden in the south and, on the northern coast and westward, a wilderness from which an Eden could be carved. To the Puritans the continent was more Promised Land than Eden. They had to pass through the wilderness before the "desert" could be transformed to a garden by a combination of God's rain and human labor. Both people and land would be blessed if the Puritans kept their covenant with God in tilling the soil and raising livestock.

This pastoral image went beyond the landscape before them to embody "a whole nexus of social, economic, and political dreams." Pathos is present in their nostalgic, unattainable dream of the garden,[4] the Simple Life. Their very quest for "the New Jerusalem" was itself the reenactment of an archetypal journey carrying them away from corruption toward redemption, both social and spiritual, with its requisite passage through the hardships of the wilderness.[5]

From our own nostalgic perspective on historical events, when the Puritans' garden wall was ultimately breeched, it was not the natural wildness that invaded but "the machine in the garden."[6] Ambivalence about the result probably resides in all of us. It was certainly characteristic of the nation's founders. Benjamin Franklin, who loved his Philadelphia, was no defender of urbanism. He distrusted industrialism and perceived agriculture as a "true source of virtue." Thomas Jefferson was "torn throughout his life by contradictory impulses." A "cultivated man who admired the cosmopolitanism of urban life," Jefferson was also devoted to the pastoral ideal. In 1800 a contemporaneous Philadelphia physician, Benjamin Rush, asserted that we are a "naturally wild animal" and never happy when taken from the woods until we return.[7]

Even novelist James Fenimore Cooper, for all his rhapsodies to lost wilderness, believed that wilderness must yield to civilization, which is the ultimate preference. The "final destiny and fulfillment of nature was for it to *become* civilization." Cooper knew his portrait of an idyllic Indian life and wilderness was overdrawn and doomed for the sake of *necessarily* converting wilderness to civilization, an assumption shared by

Cooper's hero, Natty Bumpo. Hawkeye's own closeness to wilderness was only temporary; he did not deny the "superiour importance" of the coming order.[8]

For some pioneers at least, civilization was the proud vision emergent from the "loathsome wilderness."[9] By the nineteenth century the prevailing attitude was that there was no conflict between nature and civilization, for the simple reason that nature to *be* nature *required* exploitation and civilization. For nature to fulfil its potential, wilderness was to be conquered. There was no opposition in technology; "it was the only means by which nature could become its real self."[10]

Westward Ho!

With the westward sweep of European Americans, the foundation of the relationship with nature was ownership of land, water, and cattle. Real property rights were not of immediate importance, because land and water were not initially scarce. Changes in competitive attitudes resulted in increases in numbers and kind of invading pioneers.[11]

Writings of the nineteenth century are replete with wonderment, even sermons, in awed response to the "Big Trees" and extraordinary natural areas. In 1864, while embroiled in the Civil War, Lincoln and Congress were moved to grant Yosemite to the State of California for what amounted to the world's first wilderness park. The grant was attributed to "the aura of heroic sanctity, the sense that the grove of Big Trees was some . . . living monument, a botanical pantheon." Only a generation earlier, the popular imagination had seen eastern forests as the enemy, "the habitat of the godless Indian." To replace that atavistic image with the godly pastoral one, it was necessary "that both the wilderness and the wild men be comprehensively cleared." Clearance actually equalled beauty.[12]

From the earliest days of European invasion of the American continent to the present, artists and scholars have been engaged in some form of soul-searching about the meaning of nature to the human spirit. The political process both reflected and conditioned attitudes toward nature. Whatever their personal attitudes, including ambivalence regarding nature and civilization, Jefferson, Hamilton, and the presidents following Jefferson subscribed to the principle of manifest destiny. Later, in the political era of Teddy Roosevelt and of Muir and Pinchot, Leopold expressly described the situation: "Americans were becoming increasingly ambivalent about nature. Still willing to exploit it, they at the same time sought increasingly to preserve certain aspects of it."[13]

In the decade preceding World War II, the average American saw the promises of technology and of a greater future as justification for any costs in freedom. Among the majority of people only a few were dimly aware of the costs of the journey to that juncture in terms of slavery on one hand and the treatment of Amerindians — our First Nations — on the other. Few others saw constrictions through factory systems and large corporations or indulged in some obscure misgiving with reference to the republic of Jefferson's ideal. Even fewer wondered about uncontrolled land development or what industrial growth might inflict on environmental amenities. Most took the vague threats in stride and paid little heed to any critical voices.[14]

PLACES AND A NATIONAL PSYCHE

Wilderness

How we individually perceive elements of wilderness depends on our view of the parts they have played in this continent's history. Because this continent's "wild lands have lent gravity to our national tone and public philosophy," we have been demoralized by the loss of wild rivers and wilderness. Are we a nation including wilderness or one "whose bureaucrats are in charge of its attributes, converting . . . 'beautiful, for spacious skies' to 'class II acceptable air quality'?"[15]

Too often we and our makers of policy and law have relegated wilderness and the environment to some mundane, lifeless definition in a statute whose good intentions are lost in the tedious bureaucracy of its implementation. We need to reclaim our individual relationship with the environment, wilderness, and the natural world and, out of that restored relationship, imbue with spirit our national environmental policies and laws.

Wilderness was the first victim of our "incredible economic success" on this continent. From the very beginning we destroyed wilderness as we engaged in development. It is only recently that we have come to the perception of wilderness as a scarce resource. It is only in our affluence that we begin to endorse other than economic values within wilderness. In the last two centuries abuses of capitalism have led to a societal dissatisfaction with business as usual, with special reference to the extraction of resources. We have come to a juncture where public policy is on the verge of reinforcing earlier ideas of conservation and preservation.[16]

Urbanization and the Westward Movement

There is nothing more representative of our history than the emergence of the city, but it was the decline of the city that helped drive our surrogate observers West. Promoters of the westward expansion believed that region could achieve what the East had failed in establishing "rolling farms and commercial cities existing side by side in a benevolent synthesis."[17]

A strange turnaround occurred East and West: While the Western promoters sought to build the city in the garden (*urbe in rus*), Eastern planners were seeking restoration of the garden in the city (*rus in urbe*). The underlying belief was that city life would become more meaningful were the interconnection among nature, the rustic, and the city restored. For these reforming planners the "values of Jeffersonian pastoralism could" guide the resolution of cities' "social and physical problems."[18]

Washington, D.C., and New York City are prime examples. In 1850 Andrew Jackson Downing was commissioned to design a mall along Washington's Pennsylvania Avenue, featuring a formal park with curving walks through gardens and shrubs. The mission was to enhance urban topography by providing a complementing contrast of "natural forms, carefully ordered and integrated with the city."[19]

It is difficult to believe that New York's Central Park, then also under consideration, was controversial as unnecessary. Amid the controversy raging in contemporaneous journals, a mayor was elected on a park platform holding out the park's value beyond promotion of physical health to moral implications. According to the *American Journal of Science and the Arts*, feelings would "flow in a kinder smoother channel" with more cheerfulness and happiness. Parks would "provide a release for the wellsprings of love and friendship and so lead to a tranquil and more contented society." Their omission in cities would lead to barbarism.[20]

As we moved from wild and rural settings into urban ones, our farming and forestry enterprises were freeing us from "the necessity of subsistence rural living." But the price of that freedom has been to lose touch with the land.[21] Perceptive scientists of the mid-nineteenth century were already warning that "artificially induced climatic change and the loss of species" through the "spread of Western economic development" would eventually threaten our own survival.[22]

Here are buried the roots of the contemporary environmental movement. Originating in our fascination with wilderness, the movement can be said to commence with the Romantic movement and the activist one taken up by preservationist John Muir. More recently, as disruptive human incursions have increased, an "urgent cry" has joined the other voices and latter-day Aldo Leopolds and Rachel Carsons continue to urge the maintenance of ecological balance and avoidance of outright destruction of ecosystems.[23]

The conservation movement of an earlier era has been compared with the present environmental movement: The former movement was an effort by leaders in science, technology, and government. The mission was to bring about more efficient development of resources. It was one aspect of our history of production with the emphasis on efficiency. The latter movement is more widespread and more popular. The stress is on the quality of human experience merged with the quality of the human environment. While conservation moved from top to bottom, the environmental movement came from a broader base, reaching from middle levels outward to pressure a reluctant leadership. Indeed, it would seem there are two sets of values constantly in conflict.[24]

The American West

Nothing fires our imagination in quite the same way as does the West. From one perspective the West's resources meant grasslands for wheat and cattle, forests for lumber, and the surrender of gold, silver, and copper from the mountains. The attraction was not in self-sufficiency of some Simple Life, which was something to be endured, but in transcending hardship for material gain. Among the results were overgrazing and increased erosion ultimately with the shape of landscape utterly altered.[25]

For all the reality, for more than a century, the West has been "the most strongly *imagined* section" of the country. In fact, three versions of the West can be identified: the historic, the mythic, and the image in the public mind. If myths provide meaning to the world, western myth "is a story that explains who westerners — and who Americans — are and how they should act." As the myth is accepted, it is assimilated. People act on myths, which "become the basis for actions that shape history." A striking example is Kit Carson, who upon discovering a novel of himself, capitalized "on the market the mythic Carson had created"; thus, the mythic version of the man "partially shaped the actual Carson in his image."[26]

The people of the West itself began the process of reimagination early in the process of self-styled conquest. Wild West shows were touring Europe in the nineteenth century, Pueblo ceremonials were tourist attractions at the same time the Bureau of Indian Affairs and the missionaries were out to abolish them. First novels and then film rendered the West "a defining element of American popular culture." As recently as 1958,

11 percent of all fiction published in this country were Westerns, and Hollywood was releasing a Western every week. In 1959 thirty of the prime-time television shows were Westerns, including eight of the ten most watched.[27]

Early images were of freedom, independence, the wild, the garden, the violent, and the primitive; in short, a respite from perceived evils of civilization. Many of the Western romances focused on outlaws not far removed in spirit from Robin Hood. Later, in the Western films of the sixties and seventies, we find the protagonists of *Shane, The Searchers,* and *High Noon.* In each case the Hero was making the West "safe for civilization, but it was a civilization that had no place for him." In the process, the retellings of the West were in effect a national mirror.[28]

Among the reflections are warriors, archetypal if not true archetypes. Two themes, both highly individualistic, attach to our historical and popular warriors. One is the capacity for violence for a cause. The other is motivation by conscience or circumstance, never by coercion. Reality is irrelevant; the individual warrior has been "enshrined" in our hearts. This warrior at least approaches the Hero and to some extent the Wild Man. The "lack of strong and enduring social connections . . . frees him from constraint and permits him to fight for precisely those things he is free from." Despite his high moral dimension, this warrior is so gifted in the warrior's skills as to be isolated "from the very society within which he operates" and his morality so strongly instilled as to brook no compromise. Significantly, these images of place and of archetypal persona tell us about the tellers of the tales. The reality (social misfit) is overshadowed by the myth (stalwart pioneer).[29] Their persistent popularity, albeit in altered persona, tells us of the audiences.

Wallace Stegner would bury this mythic figure but cannot because of the attraction for "the daydreaming imaginative." Moreover, "even while the cowboy myth romanticizes and falsifies western life, it says something true about western, and hence about American, character." Stegner asks why this "stereotype" has not disappeared along with the real cowboys no longer typical in the West. He answers with "the preservative" of "the visible, pervasive fact of western space." It is a space that "continues to suggest unrestricted freedom, unlimited opportunity for testings and heroisms."[30] Thus, the place and the associated persona reinforce each other.

NOTES

1. Schama, *Landscape and Memory* (1995), 574.
2. Shepard, *Man in the Landscape* (1967, 1991), 230–231.

3.   Ferkiss, *Nature, Technology, and Society* (1993), 64–65.

4.   Conron, *The American Landscape* (1974), 113, 116. See also Machor, *Urban Ideals and the Symbolic Landscape of America* (1987), 48.

5.   Machor, *Urban Ideals*, 49.

6.   Conron, *The American Landscape*, 116.

7.   Ferkiss, *Nature, Technology, and Society*, 73.

8.   Ibid., 73–74.

9.   Ibid., 77.

10.   Ibid., 73.

11.   Anderson and Leal, *Free Market Environmentalism* (1991), 27–28.

12.   Schama, *Landscape and Memory*, 191.

13.   Ferkiss, *Nature, Technology, and Society*, 93–94.

14.   Ibid., 104.

15.   McGuane, "The Spell of Wild Rivers" (1993), 62.

16.   Allin, *The Politics of Wilderness Preservation* (1982), 3, 5.

17.   Machor, *Urban Ideals*, 135, 140.

18.   Ibid., 145–146.

19.   Ibid., 147.

20.   Ibid., 147–148.

21.   Sayre, "The Rural Economy: Essential to Vermont's Character and Culture" (1992), 6.

22.   Grove, "Origins of Western Environmentalism" (1992), 47.

23.   Chapple, *Ecological Prospects* (1994), xi–xii.

24.   Hays, "From Conservation to Environmentalism," in Nash (1990), 144–145.

25.   White, *"It's Your Misfortune and None of My Own"* (1991), 236, 232.

26.   Ibid., 613, 615–617.

27.   Ibid., 613.

28.   Ibid., 621, 623, 626, 625.

29.   Donohue, *Warrior Dreams* (1994), 53.

30.   Stegner, "The Meaning of Wilderness for American Civilization," in Nash (1990), 111.

4

Repercussions: Expressions and Implications of Alienation

It has been said that human development takes the form of numerous "local experiments in creating social orders" of great variety. Each experiment, however, has also been said to work only for some of the members and that only at the expense of others. Seemingly, all the experiments have exploited nature and, while each has worked in the short term, all fail over longer intervals.[1] Our society stands accused of making a virtue of overpowering nature, especially in locating and extracting resources. One aspect of this accusation is our self-identification as "the ultimate expression of evolution," the subject of substantial doubt.[2]

Whatever our cultural attitude toward nature, individual attitudes also exist, representing a range of positions within and beyond the cultural. Oil exploration near Yellowstone National Park elicits a representative range of personal attitudes.[3] One resultant issue is how to represent the array of values, and not just one extreme, in a political setting.[4] Addressing individual perspectives is an essential step toward engaging the individual on behalf of environmental stewardship.

Today we are beyond mere pollution control. Our efforts to save wildlife and habitats "must be driven by a grander vision." The emergent objective of sustainability is to meet human need, spiritual as well as physical, without denying opportunity to future generations. That vision must guide our conduct as we come to recognize that components of the biosphere are interdependent, that humanity is a part of nature, that nature bears intrinsic value, and that both cultural and biological diversity are fundamentally natural phenomena.[5] But we must cure our ills before we can pursue the vision.

AMBIVALENCE AND ESTRANGEMENT

The logical point of departure is childhood. The life history of a child has been called an "ascension out of biological history into the world imagery of [a] particular culture." Language is the key to our "need to move . . . into a theory of the universe, to create meaning, and so to discover [ourselves] in time and space."[6] In adulthood we find ourselves in a modified, if not scarred, environment and called upon to distinguish the wounds of "greed and destructive fury" from those essential to our life and those inspired by our "love of order and beauty."[7] Those circumstances call to mind the question of what might happen to people who have lost a pastoral ethos, since our philosophy, theology, and society were founded in ancient origins of agriculture. Even more provocative is the realization that our place in nature is a narrow window opening midway between molecules and the galaxy. How can we begin to grasp all the attendent realities?[8]

It does not help that we live in a society that values a version of intellect and professionalism demanding a general aloofness we call objectivity. We all require opportunities to "get in touch with who we are and what really matters." At the same time it would behoove us to reinstate "the continuity between the conscious and the unconscious mind" through, for example, "the natural symbols" found in our dreams. We could reinforce those symbols by adding plants and animals to our immediate environments, were it not that most of us spend the greater part of our days indoors, "estranged from the vast mine of meaning, art, metaphor, and teaching" of our evolution.[9]

Historical Context

Where once our very "survival depended upon our keen attunement to the events in our environment," in recent generations our separation from nature has accelerated to the point that "our genetic and sensory evolution has not been able to keep pace with the evolution of the machine" in our "techno-oriented habitat."[10] Our current phase has been described as secular humanism, which would identify us as "the source of all goddesses and gods, the master[s] of desacralized nature, the only conscious rational being in an inanimate world." We are left to wonder if we are not "just another species thrown up by the blind forces of evolution and no doubt doomed to extinction like the dinosaurs"[11] — neither the exclusively sacred being nor in control as regent or usurper of deity.

It is Freud's (not Jung's) that is the psychology of secular humanism. Freud's grim vision is "of a lifeless, uncaring universe" yielding a "human psyche trapped in the desolation of an infinity where it finds no consolation, no remorse, no response to its need for warmth, love, and acceptance." The parallel cosmology "has nothing to warm the human spirit." It is no wonder Freud's students (including Jung) escaped so tiny a "psychic box,"[12] as we must as well.

One of the escape routes is found among our artists, who, consciously or unconsciously, have inherited an "ancient and persistent metaphorical tradition" including holy groves and sacred places.[13] And yet our own history reveals that art and science, once integrated in renditions of this continent's wildlife, are themselves a metaphor for disconnection and segregation. Natural history once embraced both zoology and the arts of literature and illustration. By the end of the nineteenth century, however, an irrevocable segregation intruded.[14]

Even our agriculture has undergone a psychic mutation. Associated technology of this century has rendered farming on a large scale not merely profitable but necessary. It was after World War I that market pressures served to sever "any connection between farm and home," at a juncture when our culture was still speaking of the two as synonymous.[15]

It is said that our damaging industrial order has become entrenched because of its "psychic entrancement" in place of preceding "mythic commitments." What must be restored, then, is the more primal mythology of ecological patterns. Thus, the mythic dimension of this new ecological age represents a deep insight into the structure and function of natural processes, whose "archetype" is ecology, including its foundation in cosmic evolution — the universe itself.[16]

By the last decades of the twentieth century we have in sequence lost sight of, track of, touch with, and respect for nature. And we have done so just as we have been able to discover "more about the detailed workings of nature than [in] all our previous millennia." In fact, perversely, "the more we learn, the more we seem to distance ourselves from the rest of life." We must recall that "[g]etting along in nature is an art, not a combat by brute force. It is more like a great, complicated game of skill."[17]

As our countryside has turned suburban and suburbs have become densely populated, we have lost our sense of place. As our mobility and its speed have increased and we have found ourselves able to choose our hometowns, we "find ourselves less and less sure of where it is we have finally arrived." While we deign to pity the songbird in winter, "the bird at least was born to the condition in which it lives. It is part of an unbroken past of this land and knows where to find itself."[18]

Expressions of Ambivalence

Ambivalence about our place in the world is the mildest form assumed by our disconnection. The unfortunate dichotomy leading from the universal Earth Mother to the urban road is reflected in "a deep psychological split" segregating emotion from intellect. The split itself is sometimes represented in "masculine and feminine stereotypes" of more than one culture.[19]

The resulting inherent ambivalence can emerge upon return from travel to a place seeming to foster the Simple Life. The traveler may return committed to self-imposed deprivation, notwithstanding the surroundings of the familiar homely comforts most of us take for granted. The commitment ultimately yields to "sensual material ease" but with a "sense of being a fallen angel," in part because we do "not know where a skillful balance" lies. Two "richly developed cultural traditions" persist in "churning and pulling" as the affected individual searches "for balance along the vast spectrum of potentials represented by these two perspectives." The internal dualism cannot be resolved. In the case of the returning traveler, it is impossible to "reject the West and embrace the East — both live within."[20]

For most of us, perhaps, the struggle to bridge two worlds[21] is encountered in the imagination, often through surrogates, especially in our popular arts. Although we share in the ambivalance and the desire for the Simple Life, we are not really prepared to encounter its realities.

Such ambivalence can actually extend into something like multiple personalities. We may each be several people. One "is or would like to be a placed person," another the opposite, displaced, "cousin not to Thoreau but to Daniel Boone, dreamer not of Walden Ponds but of far horizons, traveler not in Concord but in wild unsettled places, explorer not inward but outward." Those personalities do not share consistent needs. While "[c]omplete independence, absolute freedom of movement are exhilarating for a time," they do not necessarily wear well. The "romantic atavist" of our fantasies, while very American, "is not a full human being," but a Sasquatch.[22]

These are the reasons we need a surrogate. We cannot indulge multiple personalities and most of us cannot live either the atavistic life or in both civilization and the wilderness. That is where archetypes and contemporary versions of the Wild Man as noble savage come into play.

Outright Estrangement

The next step from ambivalence is outright estrangement. We are straying both farther and further from the natural world. Despite any sense of the ethical importance of our interactions with other living beings, with every passing year those interactions "become a smaller part of the total transaction between" us and nature.[23]

Discontinuity, separation, and denial follow. Too easily we cast about, blaming technology on one hand, Christianity on the other. And yet the sciences of ecology and transcendental romanticism, the latter with roots in the Judeo-Christian tradition, have been identified as "the two most potent forces for the preservation of nature."[24] If it is true that our "alienation from the natural world is unprecedented," a healing may mean "the difference between survival and extinction."[25]

If, indeed, as our contemporary cosmology suggests, every entity throughout time is connected to everything else, our current estrangement "is a kind of supreme evil in the universe." It is nothing less than damnation to "be locked up in a private world, to be cut off from intimacy with other beings, to be incapable of entering the joy of mutual presence."[26]

We are necessarily connected with rocks and the "nature of things." Reason may not be the bridge. To reconnect means valuing emotion, the primitive, the child, raw nature, the organic, and poetry in concert with reason, the civilized, the adult, cities, the chemical, and science. The psychological dimension of our estrangement, manifested as environmental degradation, "must be addressed if [we are] to find a graceful way to connect the mind and the world."[27]

LARGER IMPLICATIONS

Reactions and Attitudes

Our contemporary attitudes suggest hostility to the relevance of our own past. In so radically altered conditions it may seem that historical practices have become obsolete. The argument has been raised that our culture eliminates both "physical forms and social connections" supporting prior generations. We have perceived as promise a liberation from "age-old constraints." But "we are foolhardy if we base the nature of our communities on the latest technological and economic innovations while blinding ourselves to innate human needs."[28]

In the end nature and a love of nature are a part of reality, but so is alienation, even a hatred for nature, though it is certainly suicidal. For all the undeniable benefits of our culture, there is a pervasive malaise. It might be attributed to a people who lack a pastoral ethos. If that is the case, then the rural world may well be "a touchstone of renewal for the jading, corrosive drain of the city." Possibly parks and playgrounds will serve in place of pastoral settings.[29]

Our society is described in terms of one of deep imbalance, even psychosis, defined as attempting to live a lie, in this case that "we have no ethical obligation to our planetary home." Among the symptoms are violence and a fixation on sex. In the past healing was sought through restoring our deep connection with "the world we share with animal, vegetable, mineral, and all the unseen powers of the cosmos." The time has come for us to establish "a new dialogue between scientific intellect and human need."[30]

Again beginning with childhood, human specialization occurs over our prolonged postnatal infancy, but it can lead to an overspecialization approaching the pathologic. Cultural values are degraded with our "adolescent fixation on selfhood." On the other hand, to evolve a richer, more creative relationship with the natural world demands "continual expansion of cultural aims and social purposes" to release "the formative energy" of individual development. Otherwise, the risk is displacement of the "unused energy" into "the ingenuity of delinquency and crime."[31]

Even what seems a privileged childhood can suffer a deprivation in human relations. In therapy the adult may express something of the "poetics of a child's therapeutic experience in the natural world." In doing so, the adult is reliving the childhood happiness associated with ecological experiences. Often images are described as a blending of body and of world. Unfortunately, that blending lapses in adulthood into a helpless longing.[32]

Our adulthood is unlike others in which the individual learns from dreams still connected to the material world. It is more likely that our "working day begins with the dream-shattering blast of the alarm clock and a heavy dose of caffeine" to prepare us for our version of reality. Our dreams, significantly enough, are said to be relegated "to movies and television, the official, electronic fantasy-life of our culture."[33]

Another manifestation of our malaise is disaffection with place, attributed in some part to our failure to "settle anywhere" as we constantly move on to somewhere or something new. The unpleasant features of our unsettled landscape are mobile homes, chain stores, fast-food outlets, and the commercial strip. Uniformity is the mark of our transient place-lessness, and no local culture can stand up against the "forces unleashed."[34] Instead of neighborhoods we find a "collection of uncon-nected individuals."[35]

When we are so facile in replacing our homes, we ignore an emotional charge and the importance of home's familiarity to us. In fact, either "sudden destruction of a known environment" or frequent dislocation "can be fundamentally deranging."[36] When we do find ourselves in land-scapes, they "are not diffused with *our* meanings, our history or commu-nity," and it is easier for us to turn inward than to reach out to close the gap.[37] Over time we become segregated from each other[38] as well as from places or the larger natural world.

Can it be any wonder that deep psychological tensions cause us to be "pulled simultaneously away from society and toward it"? Can it come as any surprise when a person withdraws or seeks "to remold society" in an effort to eliminate or mitigate "its oppressive elements," and to replace them with others more consistent with some individual image?[39] If we avoid attachments to a thing, a person, or a place, our hearts cannot be broken. But "to withhold emotion, to restrain our passionate nature in face of a generous life just to appease our fears" is an impoverishment. When our minds rein in our hearts as the body is singing "desperately for connection," the likely result is more isolation and greater ecological disease. "Our lack of intimacy with each other is in direct proportion to our lack of intimacy with the land. We have taken our love inside and abandoned the wild." "We are a tribe of fractured individuals who can now only celebrate remnants of wildness." "But what kind of impoverish-ment is this to withold emotion, to restrain our passionate nature in face of a generous life just to appease our fears?"[40]

The Healing Process

Carl Jung writes of *unus mundus,* an "ineffable unity" at "the core of mystical illumination." Significantly, Jungian psychologists ask whether the "relationship of particle to wave in quantum mechanics might, at least symbolically, express the complementary relationship between the conscious and unconscious mind." The parallel psychiatry opens doors for "religiously estranged Westerners" that may represent a contemporary "quest for illumination through knowledge of the mysteries."[41]

Psychological reviews in the eighties criticized the tendency toward "fragmentary, unrelated research" projects and warned of the hazards in the lack of a theoretical approach addressing the intricacies of the interaction between human and landscape. Stephen Kaplan has since explored environmental preference, incorporating evolutionary considerations. His thesis calls for a "new and larger role for aesthetics, broadly construed, as a central force in human experience" and behavior.[42]

For most of us, myth is devalued and disempowered as untruth. We have lost that "great reservoir of imagery designed to transform consciousnessness and provide a framework for society." Our society is marred and fractured by its lack of some "commonly held belief system, energized by archetypal imagery."[43] Even Jung accepts as irreversible "our estrangement from forest and sea, rivers and mountains, and from our brother and sister creatures."[44]

The recent popularity of the work of Joseph Campbell may be testimony to the popular attraction for "combing through the world's forgotten lore" in seeking "to salvage things of living value." The risk here is that this approach to enlightenment can, like all other endeavors taken up by scholars, "become oppressively pedantic."[45]

Religion, too, has faded from our lives, removed to the fringes until a crisis looms. In our past, religion "celebrated the tasks of a working day, the seasonal unfolding of the year, and the plateaus of personal development." In short, religion pervaded life, culturally as well as individually. Lost, too, are rites of initiation, taking the child through death and rebirth into adulthood with its privileges and responsibilities. In that process naive realism is exploded to reveal "that things are not simply what they appear to be, that one-dimensional literalism is" childish. The one who does not grow beyond it must "despair of a life rich in meaning and worth."[46]

It is time to reclaim what we have lost. We would again "feel the presence of other living beings and of the natural world" as a part of us, with intertwined destinies.[47]

Our profound sense of loss can be evoked by essays that may elicit "vivid childhood memories of contact with wild nature." There appear to be "common ways in which wildness — even in its simplest forms — can nourish a lasting attachment to the earth, and, in turn, nurture self-esteem." In short, children *need* wildness.[48] And many adults write evocatively of momentary connections to the natural world.[49]

It is reported that in recent years both women and men have experienced unusually frequent dreams of injured animals. The coincidence with devastation of wilderness is noted as is the dreams' reflection of "deep lacerations in the collective unconscious regarding the loss of the instinctual life."[50]

Thomas Berry reflects on our need for the healing of restoration with nature in what once might have been deemed banalities — "the world of dawn and sunset and starlight, the world of soil and sunshine, of meadow and woodland, . . . of wildlife dwelling around us, of the river and its well-being." That these are not banal, but a richness "some of us are discovering for the first time" is testament to our need for healing. "Here, in this intimate presence to the valley in all its vitality, we receive those larger intuitions that lead us to dance and sing, intuitions that activate our imaginative powers in their most creative functions." Once again place is fundamental: out of each place "a unique shaping of life takes place, a community," integrated with components of geology as well as biology and humanity.[51]

Today, self-styled ecotherapists are emerging: "Everyone knows a stroll in the woods can be good for the soul." Some therapists "see it as a key to mental health." This practice, however, lurks on the fringes as untested with the potential for quackery. But ecopsychology is taught at Harvard Medical School. Treatment regimes range from incorporating shamanism to wilderness treatment, simply encouraging a closeness to nature or gardening.[52]

Those who study contemporary women also speak of our longing for our own wild kind with its promise. Vicariously or in our dreams we experience "a wondrous wild world" to which we once belonged. The "memory of it is a beacon that guides us toward what we belong to . . . for the rest of our lives." Here, too, nature is perceived as a source of "powerful and straightforward" healing. There is therapy all around us.[53]

Animals are a vital link to interconnection, expressed today in wistful rumors and anecdotes of a direct "mind-link" between us and dolphins. The longing for linkage is not limited to dolphins. It somehow awakens "the future primitive" in us. Perhaps this is where we will be when we become "more civilized." A "meditational interface between human and nature offers a genuine bridge between" science and "the ancient shamanic-mythic view." The process is a personal one promising a cultural interconnectedness without negating the products of our brains.[54]

At the very least, nature is seen as restorative. D. H. Lawrence expressed it as leaving "the glass bottles of our ego" to escape into the forest, where "we shall shiver" but where "things will happen to us" with "cool, unlying life" rushing in to embue us with new power. We will laugh and "institutions will curl up/like burnit paper."[55]

We have reached a juncture in our culture where the forest is no longer to be avoided as dark, forbidding, sinister. Swamplands have become wetlands, valued natural systems. Predators bear a positive image. No longer is the desert forbidding, but a place of wonder and beauty. All of these transformations have come about in the years since World War II.[56] Nature is perceived as "our widest home" including "ancestral habitats" once "shared with countless kinds of animals," as "the central theater of life." Now it is more "a hideaway where" we go "to rest and recharge after a hard stint in an urban or suburban arena."[57]

NOTES

1. Brunner et al., *Corporations and the Environment* (1981), 52.
2. Mander, *In the Absence of the Sacred* (1991), 6–7.
3. See, e.g., Anderson and Leal, *Free Market Environmentalism* (1991), 82; Sagoff, "Setting America or the Concept of Place in Environmental Ethics" (1992), 357–358.
4. Anderson and Leal, *Free Market*, 82.
5. Berle, "Toward a Grander Vision," (1995), 6.
6. Cobb, *The Ecology of Imagination in Childhood* (1977), 50, 51.
7. Jackson, *A Sense of Time, a Sense of Place* (1994), vii.
8. Shepard, *Man in the Landscape* (1967, 1991), xxxv–xxxvii.
9. Gallagher, *The Power of Place* (1993), 203.
10. Mander, *In the Absence of the Sacred*, 34.
11. Sheldrake, *The Rebirth of Nature* (1991), 33.
12. Ibid., 58, 59.
13. Schama, *Landscape and Memory* (1995), 207.
14. Blum, *Picturing Nature* (1993), 318.
15. White, *"It's Your Misfortune and None of My Own"* (1991), 437.
16. Berry, *The Dream of the Earth* (1988), 32–34.
17. Thomas, *The Fragile Species* (1992), 100, 123.
18. Finch, *The Primal Place* (1983), 5.
19. Roszak, *The Voice of the Earth* (1992), 138.
20. Elgin, *Voluntary Simplicity* (1981), 16–17.
21. Ibid., 22.
22. Stegner, "The Meaning of Wilderness for American Civilization," in Nash (1990), 200.
23. Shepard, *Man in the Landscape*, 204.
24. Ibid., 220.
25. Orr, *Ecological Literacy* (1992), 17.
26. Swimme and Berry, *The Universe Story* (1992), 77–78.
27. Roszak, *The Voice of the Earth*, 41–43.
28. Langdon, *A Better Place to Live* (1994), xiii–xiv.
29. Shepard, *Man in the Landscape*, xxxviii, xxxv–xxxvi.
30. Roszak, *The Voice of the Earth*, 13–14, 17.
31. Cobb, *The Ecology of Imagination in Childhood*, 110–111.

32. Ibid., 75-77.

33. Roszak, *The Voice of the Earth*, 81.

34. Sagoff, "Settling America or the Concept of Place in Environmental Ethics," 354.

35. Langdon, *A Better Place*, 18–19.

36. Tall, "Dwelling: Making Peace with Space and Place" (1995), 16. Tall cites statistics to the effect that 20 to 30 percent of Americans move every year. The average person moves fourteen times over a lifetime. We seem to favor moving up, moving on.

37. Ibid., 20. These excerpts were taken from *From Where We Stand: Recovering a Sense of Place,* published in 1993.

38. Langdon, *A Better Place*, 64.

39. Machor, *Urban Ideals and the Symbolic Landscape of America* (1987), 17.

40. Williams, *An Unspoken Hunger* (1994), 63, 64, 65.

41. Roszak, *The Voice of the Earth*, 61–62.

42. Kaplan, "Aesthetics, Affect, and Cognition: Environmental Preference from an Evolutionary Perspective" (1987), 4–5, 7.

43. Broadhurst, "The Subtle Power of Myth" (1995), 12-13.

44. Roszak, *The Voice of the Earth*, 63.

45. Ibid., 63.

46. Gill, "Disenchantment," in Dooling and Jordan-Smith (1989), 106, 113.

47. McDaniel, *Earth, Sky, God & Mortals* (1990), 29. See also editorial, *Orion* (Summer 1995), 2.

48. Nabhan and Trimble, *The Geography of Childhood* (1994), xii–xiii.

49. See, e.g., Ibid., 35–37, 40; Tuan, *Passing Strange and Wonderful* (1993), 57; Trimble, "A Land of One's Own," in Nabhan and Trimble (1994), 61.

50. Estés, *Women Who Run with the Wolves* (1992), 276.

51. Berry, *The Dream of the Earth*, 176–177.

52. Aeppel, "Ecotherapists Explore the Green Side of Feeling Blue" (1995), B1, B5.

53. Estés, *Women Who Run with the Wolves*, 189–190.

54. Nollman, *Spiritual Ecology* (1990), 180.

55. Lawrence, "Escape," in La Chappelle (1988), 270.

56. Hays, "From Conservation to Environmentalism," in Nash (1990), 149.

57. Mosher, *Songs of the North* (1987), v (Edward Hoagland regarding Penguin Nature Library).

II

IMAGINATION AND THE ARTS: WHERE WE SAW OURSELVES

5

Emergence of the Simple Life

Nature inspires all of humankind. Of that there can be no doubt. Our arts — music, dance, paintings, and literature — celebrate and imitate the beauty of nature. The wonders of nature are said to be "universal in effect and timeless in appeal."[1] Throughout Western traditions such impressions have been expressed by celebrated poets, "[o]ne touch of nature makes the whole world kin,"[2] and have not been lost on renowned scientists; "no one can stand in these solitudes unmoved, and not feel that there is more in man than the mere breath of his body."[3]

There was a time when it was possible *and respectable* for any "cultured" person to be attuned to the combined arts and letters of Western culture. Indeed, the arts intermingled freely with the sciences, and philosophy and theology were an integral part of a cultural whole. If there were disconnections and estrangements among the lines of human thought or between humankind and the natural world, they were not at the forefront of cultural expression.

Mythology and lore evolve alongside natural history and metaphysics, and at times the four are not readily distinguishable. Nevertheless, each has a role in informing the culture and the individual. The major themes of human relationships to the natural world emerge early in the Western tradition. Homeric hymns were sung to Earth, as the mother of all, and honored the eldest of the gods as the nourisher of all the things of the world.[4]

In recent times, *USA Weekend* has distributed a special publication acknowledging the existence of "ancient guidelines for human behavior which are found in the teachings of the religions of the world and which are the conditions for a sustainable world order." This publication makes reference to the symptoms and multiple levels of human estrangement not only from the natural world but also from each other.[5] And yet we have evidence reaffirming the earlier poets and scientists in such seemingly disparate phenomena as modern drumbeats unmistakably recalling sexual rhythms and even more ubiquitous primal patterns and Christian hymns, among which *America the Beautiful* stands representative.

In this context it has been asserted that the literature of every time denies "as well as affirms the ideals of that period." Sixteenth-century literature was suggesting that civilization is corrupt.[6] The Simple Life is Western civilization's metaphor for a preferred cultural relationship with nature. The Simple Life takes three basic forms: pastoral, atavistic, and utopian. Closely linked with these images of relationships to place are images of metaphoric — or surrogate — persona. Moving among the images of the pastoral, atavistic, and utopian are the Hero, the Wild Man, and the noble savage — who combines the two archetypes.[7] All of these metaphors of our relationship to nature stand for our cultural and individual connections with and estrangements from the natural world.

MANIFESTATIONS OF THE SIMPLE LIFE

Pastoral

The pastoral version of the Simple Life is literally the garden (or Arcadia), found in both literature and art as one of the first images of nature. The garden has been called the "purest microcosmic expression of the man-nature relationship."[8] Over the generations from the classical world to our own, gardens have been sanctuaries from "the evil of the world," and whenever the garden wall is removed there is a faith that all of nature can serve that same end.[9]

Arcadia was a province of ancient Greece renowned for its sweet air and well-tempered minds, more recently described as a kind of England in perpetual Maytime.[10] During the Renaissance the theme took on a familiar form. Arcadia represents a golden age of plenty with rich fields. Iron, war, and destruction are all absent, but in this setting of "impossible sweetness" love is thwarted and all is not necessarily "birdsong, wild honey, and nosegays in the moonlight." Darker emotions abide as well.[11]

Pastoral poetry is of remote origins, perhaps the spontaneous songs with which shepherds whiled away their time. But pastoral literature depends on urban communities for the inspiration of contrast. The founder of this poetry is Theocritus, followed by Aristophanes, whose works were the first expression of a literary past. By his time, city dwellers were both conscious of the different ways of life and sufficiently developed artistically to put that consciousness into verse.[12]

In fact, it appears that whenever cities reached some minimum size and complexity, at least the gentry turned to the farm and country living. The trend may have emerged in the time of the Greek and Roman philosophers, but it has continued to the present.[13] That the trend is not unique to Western traditions will become apparent in later chapters.

In *The Daphnis Song,* Theocritus's Idyll I, a goatherd and shepherd praise each other's music with comparisons to trees and flowing waters. As Daphnis languishes, he bids farewell to the wild creatures and to the dells, groves, and woodlands. Jackals, wolves, and lions, as well as domestic beasts, weep for him, and Daphnis calls on the woodland plants to change their nature on his behalf.[14]

The Roman gentry were not content with poetry but retired to a country house for its "pastoral surroundings, horticulture, and unhurried quiet." The design was concentric, with the house in the center of the garden and a park beyond. All rooms opened onto a terrace and were themselves gardenlike.[15]

Italian villa gardens of the sixteenth century were artfully designed with an assumption of the visitor's familiarity with the classics and pagan myths represented. Often the allusions progressed from the wild to the civilized or in reverse toward a primal source. Similarly, such gardens might be sited at a "symbolic boundary between the visible and invisible worlds,"[16] suggesting that Europeans of that era still connected to the spiritual realm, often deemed lost to them and to their descendants in the contemporary United States.

The popularity of the garden was reflected in the landscape gardening books of the eighteenth century, some published as early as 1700. Significantly, these gardens, extolled as "rude wildernesses," were in fact landscaped versions of wilderness, an indication of how close our mental images of the Simple Life are. Indeed, the prospect of such gardens extended in an unbroken view to the landscapes beyond.[17] Where the garden wall was removed, earth itself became an extension of the garden. Rousseau described the English park as a "wilderness garden," in which "naturalness . . . nurtured an esthetic which would be directed to the actual wilderness."[18] By the end of the century there were, in modern terms, arcadian theme parks with mechanical volcanoes, labyrinthine caves, rivers Styx, and even artificial thunderstorms.[19]

Today's closest approximations to pastoral reality are found in wild lands of the savannah, steppe, and scrub; "precisely the type" of lands our predecessor humans and protohumans of the Paleolithic probably hunted.[20] Thus, from the garden, the Simple Life leads us into the wilderness and the atavistic.

Atavistic

Ironically, it is agriculture itself that served to eliminate wild nature and to dominate human landscapes, thus leading to appreciation of wild nature.[21] Atavism, however, is not so large a step from the pastoral. Atavism, a return to a more primitive state also known as primitivism, is expressed in numerous ways. Because humankind is a part of it, nature is central to all of them. Observers Lovejoy and Boas have listed more than sixty versions of atavism in antiquity alone.[22]

This version of the Simple Life is said to invite us to be ourselves, to vent natural desires, ordinarily constrained through civilization, and thus enter a realm more naturally ours. Atavism is a vision of something we have lost and reassures us of its latent presence despite our corruption by civilization.[23]

Fire is particularly evocative of atavism: "For millennia men squatted around the warming, . . . dancing fire, fascinated, comforted, transported, mesmerized by light in the dark, the colors, odors, sounds." A small fire evokes "the hushed memory of the past"and "the unknown and the mystery of life are admitted to the communal mind." In short, "fire has an incredible hold on the modern psyche." There is "an expectancy, as though some secret were about to offer itself as a clue to the greater reality."[24]

For Europeans of the sixteenth century, animals were separate and deemed resources for "human society and economy." Although beasts were ranked lower than humankind, they did carry an allegorical role. They became romanticized representatives of the primitive, especially with the closure of lands once common as reserves for the privileged classes.[25] A member of the animal kingdom transcending cultures is the dolphin. From the early art of the Western tradition to the present, the dolphin is no less than "the talismanic beast for a safe and blessed journey across water, often from mortal to the immortal realm."[26]

Utopian

According to Simon Schama, the myth of the mountain utopia was reinvented in the Europe of the eighteenth century, as a part of that time's "obsession with primitive virtue." Representative of this image of the Simple Life is the 1732 poem, *Die Alpen,* by Albrecht von Haller, which was "rapidly translated into all the European languages" in countless editions. This utopia, protected from such lowland vices as greed and luxury, was made possible by the mountain's "blessed barrier." Here it was possible to drink "the cold, clear water that gushed from mountain brooks, inhale the pure alpine air untainted by the stinking miasma of metropolitan life." The resident Alpine peasant is a kind of precursor for the wild man or noble savage in his rustic timber home and fed and clothed by his habitat — "governed by the laws of nature, not the legacies of Rome."[27]

It is important to remember that these images of the pastoral, the atavistic, and utopian are just that — images. They are not necessarily related to any reality. In fact, it is far more likely that the more deeply one is immersed in either the garden or the wild, the *less* enamored one is of that way of life. Harsh realities intrude upon the ideal invented by imagination. Yet the pastoral and the atavastic are "a collection of images about the world which need not be practical . . . to be valid. [Urbanites], who want nothing to do with the smell of . . . dung, . . . incorporate appropriate elements of pastoral mythology in their philosophy."[28] While peoples we consider primal or wild do not dote as we do on their "simpler" worlds, they are not indifferent to them. Instead, these peoples are "profoundly in transaction with nature" and their environment is an intrinsic blend of religion, mythos, and landscape.[29]

Directed to whatever place moves us, the aesthetic goes beyond the material into the metaphysical, the supranatural, and the supernatural. There is an element of the magical.[30] Who better to carry us into those realms than gods and heroes?

GODS, HEROES, AND NOBLE SAVAGES

God and Wild Man

Representative of classical works relevant here is *The Bacchae* of Euripides,[31] dramatizing the myth of Dionysus, son of Zeus and the mortal Semele, princess of Thebes. As the drama opens, Dionysus has been rendered mortal to restore his mother's name and reclaim his own godhead. He has enrobed his mother's tomb with *his* vines. Dionysus himself wears a wild fawn skin about his shoulders and carries a javelin wound with ivy. He has punished the betrayers of Semele, her sisters, by making them his followers: "From their chambers their home is made the 'bare mountain side.' With hearts aflame they live among roofless rocks and shadowy pine trees green." The chorus of Eastern women who are followers of Dionysus are clothed in light of sunrise with white robes, ivy-bound hair, and fawn skins. Some carry sacred wands, trees ringed with ivy. All are exhorted to worship on the mountain and are called up into the hills.[32]

Every image reveals Dionysus as something of a noble savage, if not the Wild Man. He is of the wild and calls his faithful and those he has condemned into his wilderness. Euripedes' poetry, even in translation, is extraordinarily evocative. The modern reader, at least, might well long for his world, not of debauchery but of the wild freedom of a colt "who runs by a river . . . when the heart of him sings."[33]

Dionysus is initially found on a rock seat with "ward o'er birds and wonders." He is later caught and brought before the king, whereupon the commander of guards reinforces the wildness of Dionysus in announcing the quest finished with prey caught after a swift chase, "and this wild thing/Most tame; yet never flinched, nor thought to flee." While a captive of the king, Dionysus is held in the stable. He calls to his followers and then upon the "Chained Earthquake" to awaken, which it does and shakes the castle. Dionysus next calls upon "Lightning's eye" and "the fire that sleeps" before he emerges unbound from the castle. Freedom is expressed by the chorus as well, who sing of the dew on their throats and the stream of wind in their hair and of the fawn fleeing to the greenwood "[a]lone in the grass and the loveliness," of the "[l]eap of the hunted, . . . /Beyond the snares and deadly press."[34]

The drama goes on, with the excesses of Dionysus's wild life contrasted to the civilized ways of the king. The reader cannot help but be drawn, as are the captives, to the Simple Life of the woodland. Neither, however, would choose to be lost forever to the wildness. To be so lost is to lose the responsible part of oneself. Dionysus, then, is an early noble savage, caught between two worlds, who evokes in others the wildness counterbalanced against the civilized. He has recently been described as a latecomer to Olympus who "never ceased to be a stranger there."[35] We, too, are strangers, sometimes to one world and sometimes to another.

Noble Savages and Wild Men

Tacitus, for one, contrasted Teutonic life with that of the Roman Empire of the first century A.D., when the Empire was entering its decline. To Tacitus these barbarians represented a simpler life and were at once more virile and more virtuous than Romans. Among the qualities he assigned the barbarians with obvious approval are courage and hardiness, hospitality, democracy, and chastity as well as a freedom from the complexities of a luxurious Roman life, and a lack of avarice. And they were good mothers![36]

This array of distinctions also represents lineages of noble savages and Wild Men that remain with us today. In fact, it is the role of the Wild Man to comment, usually in negative terms, on the civilization that creates him. Tacitus anticipates those lineages in perceiving "a noble and virtuous race living in a state of savage simplicity" and in contrasting by implication their "moral excellence" "to the vices . . . sapping the strength of the more complex and pretentious Roman civilization."[37]

Whether philosophers acknowledge them or not, each generation has had its own version of "primitive innocence." Even in our own industrial age there are figures who challenge the rightness and rationality of civilization. From Oroonoko[38] in the seventeenth century, to Uncas and Friday of the nineteenth, to Tarzan and Mr. Spock of the twentieth these inventions have been widely popular reflections of our fascination with virtues we may suspect our civilization actually endangers.[39]

To Rousseau the state, science, and culture all serve to corrupt the human estate, while a natural existence provides happiness and well-being. It was he who engendered the noble savage and proclaimed the savage life virtuous. In such primitivism "wild nature was idealized as an oasis free of the ills of civilization, a retreat to which the harried and battered, the suppressed or oppressed, might turn for relief."[40]

NOTES

1. Mattingly, *Songs of the Earth* (1995), first and second pages of introduction (unpaginated).

2. Shakespeare, in Mattingly (unpaginated).

3. Charles Darwin with reference to primeval forests, in ibid. (unpaginated).

4. Oates and Murphy, *Greek Literature in Translation* (1944), 905. Homeric hymns are associated with the eighth to the fifth century B.C., and are of the style and meter of Homer.

5. "Towards a Global Ethic: An Initial Declaration," distributed by *USA Weekend* as a public service, produced by 1993 Parliament of the World's Religions (consisting of more than two hundred scholars and theologians), 1, 2.

6. Fairchild, *The Noble Savage* (1928), 18.

7. Although it is not always easy to sort archetypes from the archetypal and other, lesser images, upper case is reserved for archetypes in the Jungian sense.

8. Shepard, *Man in the Landscape* (1967, 1991), Plate 5, between 170 and 171.

9. Ibid., 88.

10. Schama, *Landscape and Memory* (1995), 531.

11. Ibid.; Jacopo Sannazaro's *Arcadia* was published in Venice in 1519.

12. Oates and Murphy, *Greek Literature*, 927.

13. Tuan, *Passing Strange and Wonderful* (1993), 111.

14. Ibid., 927–931.

15. Shepard, *Man in the Landscape*, 68.

16. Schama, *Landscape and Memory*, 275, 276.

17. Ibid., 539.

18. Shepard, *Man in the Landscape*, 88. With reference to English gardens in this context, see Ferkiss (1993), 56–57.

19. Schama, *Landscape and Memory*, 541–542.

20. Shepard, *Man in the Landscape*, 54.

21. Blum, *Picturing Nature* (1993), 5.

22. Hermundsgård, *Child of the Earth* (1989), 94.

23. White, "It's Your Misfortune and None of My Own" (1991), 26.

24. Shepard, *Man in the Landscape*, 55.

25. Blum, *Picturing Nature*, 5.

26. Schama, *Landscape and Memory*, 275.

27. Ibid., 479–480.

28. Shepard, *Man in the Landscape*, 50.

29. Ibid., 53.

30. Ibid., 29.

31. As translated by Gilbert Murray in Oates and Murphy, *Greek Literature*, 341.

32. Ibid., 343, 344–345.

33. Ibid., 347.

34. Ibid., 352, 354, 359, 367.

35. Boorstin, *The Creators* (1992), 202.

36. Fairchild, *The Noble Savage*, 4–5.

37. Ibid., 5.

38. Oroonoko was the creation of Aphra Behn in the 1870s. As a royal slave from a noble civilization, he was not a "true child of nature," but he is very much

the predecessor of our contemporary vision of the noble savage. See Ibid., 35–36, 37, 40.

39.   Roszak, *The Voice of the Earth* (1992), 223.
40.   Oelschlaeger, *The Idea of Wilderness* (1991), 111.

6

Sciences, Art, and Literature

NATURAL HISTORY AND ART

The nascent scientific revolution afforded our culture a new knowledge of the intricacies of and the vast powers at work in nature. At least initially, natural law was no longer external; philosophers and aestheticians perceived a "moral and aesthetic norm" in nature. Where scientists and philosophers led, artists followed — all expressing cultural relationships to landscape and natural history on our behalf.[1] For some observers, there was a compelling juxtaposition of the natural and social orders. While the wild and the civilized may well be in conflict, the wild can inform society, and nature at large actually enjoys the "more perfect organization."[2]

In America, for writers and artists alike, there is a spiritual connection with the land. Their renditions are of an ideal place of the imagination too readily dismissed as mere sentimentality and escapism, simplistic and banal. These works should be viewed as bearing a psychic element almost certainly connected to our collective unconscious and some of its archetypes. Just as the scientist exploring the landscape works with biological systems, the poet turns to "painterly impressions and mental associations" and the painter to line, texture, and light.[3]

As late as the eighteenth century, the three observers of landscape — scientific, artistic, deistic — were still sharing a sense of harmony, beauty, and the sublime. The natural world was a revelation to each of them, and the ideas inspired by landscape were accessible "to anyone who would look for them in galaxies or in earthly landscapes — especially in wilderness landscapes, untouched by" humankind. The sublime was said to emerge from a combination of obscurity (in darkness and mists), power (of a storm or waterfall), and privation (imposed by darkness, silence, or solitude). Any number of combinations of natural phenomena, "consecrated as Romantic icons," serve the artist in conveying so emotional a response to the world.[4]

In this context the work of Frederick Remington comes to mind. Contemporary artistic descendants of the eighteenth and nineteenth century observers are Sabra Field and Bev Doolittle, whose works could hardly be more diverse. And yet each conveys a sublime relationship to landscape, artistic yet accessible.[5] Another genre altogether, but one worthy of critical attention, is that of the "spacescape" artist who renders either real galaxies, stars, and planets or those imagined by authors of science fiction and fantasy.

Landscape

In Writings

Something of our historical ambivalence toward nature and civilization is discernible in landscape art. It is worth noting that from the earliest writings on landscape, the literature represents less a scene than a total immersion. For such writers landscape is a milieu and a vital part of the person.[6]

There is an early seventeenth-century ode to "the Virginian Voyage" based on another's report of the first voyage to that land in 1584. The poet tells of "the luscious smell," "that delicious land," of the flow of the sea and of the clear wind, causing "hearts to swell."[7]

Such attitudes were echoed in sermons as well as poetry. Boston Unitarian and "natural missionary" Starr King described the Sierra Nevada as the "visible face of divinity" and the "purest American habitat." Our generation may be at a loss in so combining spiritual with material outlooks, but our recent ancestors had not yet made the split. In fact, some of them saw the Sequoias as a sacred part of "America's own natural temple" and were intrigued by the realization that these trees are contemporaries of Christ.[8]

Closer to contemporary nature writing is William Bartran's *Travels,* describing his 1776 experience of Georgia's Altamaha River. Bartran writes that he spread his "skins and blanket by my cheerful fire, under the protecting shade of the hospitable Live-oak, and reclined my head on my hard but healthy couch." In this idyllic setting he "listened undisturbed" to "the divine hymns" of songbirds in groves, "whilst the softly whispering breezes faintly died away." With sunset and moonrise "how melodious is the . . . mock-bird! The groves resound the unceasing cries of the whip-poor-will." And "at length" a "universal silence prevails."[9]

In the Visual Arts

Two landscape schools of visual art stand out: the Hudson River School of Romantics and the artists specializing in the frontier and depictions of Amerindians. Representative of the two schools are Thomas Cole and George Catlin, respectively. Cole actually idealized that juxtaposition between wild and civilized. His work glorifies nature while still finding the natural world menacing. From a modern perspective he "clearly sensed something was wrong with America's assault on nature, and that the country might someday have to pay a terrible price for its headlong pride."[10]

Natural History

Early natural history in America, mixing science, art, and theology, extended beyond the landscape to embrace its flora and fauna. Illustrations by naturalists represented one expression of our society's perceptions of animals and their meanings, in psychological terms. The important assumption behind publication of these illustrations is that of a "growing literate public" for them.[11]

In the early nineteenth century, naturalist Alexander Wilson resorted to anecdotes attesting to his firsthand experience with the subjects of his study and engaged in "extended lyric verse when mere prose failed to convey his feeling for nature." Wilson elected to entertain as he instructed, in the belief that work promoting "an appreciation and knowledge of nature" contributed to national progress.[12] Wilson's contemporary, entomologist Thomas Say, was emphasizing the science of taxonomy and distancing himself from the text. He did, nevertheless, provide a verse on the title page. Here was an early version of ambivalence: The naturalist should not intrude on the science, yet Say did afford "special attention to visual elegance" in his 1817 *American Entomology*. Significantly, Say was concerned with broadening the audience for his work, and the critics shared his belief in "patronage of natural history by a nonspecialist readership." At the time those promoting American culture held science to be among the contributors to "the nation's developing arts and letters." Natural history was literally a component of human history.[13]

On the twofold premise that nature expresses a divine plan and that human efforts to comprehend nature emulate creation's own intelligence, naturalist-artist Louis Agassiz asserted that naturalist-artists bore a responsibility for illustrating "divine thought expressed in individual organisms." Agassiz is the author of the ten-volume *Contributions to the Natural History of the United States,* for which nearly a third of a million was raised through some twenty-five hundred advance subscriptions. An anti-Darwinist, Agassiz revealed in his taxonomy "plan, structure, design, pattern, and symmetry." Both fundamental anatomy and "the minutest ornamentation all represented . . . thoughts of God." Another critic of Darwin, John Strong Newberry, found the theories to invoke "beauty and the mysterious origins of life and of human consciousness as evidence of divine creation."[14]

The year 1830 saw the inauguration of the monthly *Cabinet of Natural History and American Rural Sports.* Entertainment was its principal mission, but it would incidentally instruct. There were companion periodicals, including *American Natural History* and *American Journal of Science and Arts.* Although some offerings were deemed taxonomic and others sentimental, many readers clearly saw natural history as a form of entertainment. Perhaps America's best-known artist-naturalist John James Audubon avoided any "genre division" by firmly "anchoring his enterprises in the domain of fine art." Indeed, his life's work was staked on its broad appeal and his attitude was one of observer as participant.[15]

In this era, George Perkins Marsh, in many ways a predecessor of today's interdisciplinary environmental professional, was observing nature in the capacity of lawyer and what we would now call a conservation biologist (ecologist). For all his attention to ecological detail in the context of anthropogenic perturbations, Marsh was not without a poetic side: "The bubbling brook, the trees, the flowers, the wild animals were to me persons, not things," many men "would find it hard to make out as good a claim to *personality* as a respectable oak can establish."[16]

A spirit kindred to the landscape- and naturalist-artists is that of Henry David Thoreau. Thoreau's *Walden* has been described as a mosaic in which are mixed philosophy, natural history, geology, folklore, archeology, economics, politics, education, and more. The man lived his subject and, as is the case with each of us, his discipline was as broad as his own imagination. Neither he nor we are restricted to some single "academic pigeonhole." *Walden* is, in fact, "a model" for a "possible unity" of "personhood, pedagogy, and place."[17]

LITERARY MOVEMENTS

Romantics

Romantic naturalism has been defined as a search for the supernatural within the natural for "an emotionally satisfying fusion of the real and unreal, the obvious and the mysterious."[18] Romanticism is identified with a celebration of nature, especially the wildness in nature. Paralleling this celebration is a refutation of society (civilization) as unnatural and corrupting, which leads us to Rousseau's solitary, instinctive communion with nature.[19]

Elements of the romantic movement were discernible in 1730, and the movement was fully developed by 1790.[20] An earlier representation of this literary genre is found in Cervantes's first extended non-theatrical work, *La Galatea,* of 1585. Described as the "familiar escapist genre of the day," this is a pastoral romance of shepherds and shepherdesses in an idyllic countryside. Eighteenth-century cities were denigrated as confining and smoky while groves, valleys, and even caves were extolled as warm and loved. According to such writers, while primeval man suffered his troubles, ours are worse because they are unnatural and brought on by our own corrupting desires.[21]

Percy Bysshe Shelley, one of England's romantic or transcendental poets, expressed in a letter of 1811 a longing for a time when men lived in accord with Nature and Reason, "in consonance with virtue." To him the barriers to such accord were religion and policy and their respective establishments. With maturation in personal philosophy, Shelley conceived of a "spiritual chain" that sounds surprisingly like an amalgam of metaphysical attitudes from a variety of cultures with contemporary ecology. Shelley's spiritual chain "links all creatures together" and "places each individual soul in actual contact with the soul of the universe." There is poetry in his essay, *On Love,* with its references to the motion of leaves in spring, eloquence in the wind, and melody in brooks and rustling reeds, all of "which, by their inconceivable relation to something within the soul, awakens the spirit to a dance of breathless rapture, and brings tears of mysterious tenderness to the eyes."[22]

Of particular significance here, the nineteenth-century English romantic poets saw an immanent God whose will was being worked out by humankind. They wanted us to *re*-establish an earlier intimate relationship with nature as a source of joy and inspiration. While elements of nature itself could be appreciated through the senses, any higher emotion required "a contemplative mind that could enter into direct communion" with nature.[23]

Christian faith remains at least an implicit component of romantic poetry. In the poetry of John Keats, for example, deity goes beyond creation and becomes a mighty, unifying intelligence, an identity toward which the human mind aspires for perfection.[24] Nature revealed no Godhead to Lord Byron, nor was it identified with a Creator. But for this poet nature is a divine work blending all beauties and evoking "a pure passion of love and reverence" leading the beholder to condemn the mundane as insignificant in face of "sublime physical phenomena." A "Personal God" stirs the soul and brings one into the acquaintance with "the Eternal harmony of the universe" leading to the perception of "everything in nature . . . encircled with the girdle of beauty," a tangible "spirit interpenetrating all." From Shelley: "Nature's life and beauty move in an ever-repeated order, revealing the mighty, unseen, and eternal power that works behind her." In prose, an excerpt from Shelley's *Adonais* speaks of Nature's living Spirit breathing "complete accord and unity, which man must learn, . . . if he wants to enjoy in full the beauty, power and majesty of his being."[25]

Psychology is yet another discipline informing environmental stewardship and one whose elements influenced the romantic poets. The works of William Wordsworth reflect psychology along with philosophy without the jargon. To Wordsworth nature speaks to us everywhere, from within as well as from the outside, and "arrested his attention" from early childhood. He clearly articulates a sense of connection with place extending his "sympathies" "to all the objects around him." Nature's "serenity and peace" became "a permanent source of joy . . . while her majesty and grandeur impressed him with a sense of reverence, even awe."[26]

Transcendental

Closely akin to the poetry of the romantics are transcendental writings. Transcendentalism is attributed to the English romantics as preachers of a simple, naive religion free of external authority or edict. Fulfillment resided in knowing oneself and how to act naturally, guided by the code provided by the laws of nature. In fact, just as natural law was discovered through science and spirituality, such a law lay within the person. Transcendentalism might exclude pantheism, but it attributes to nature a certain sanctity "as the most clear manifestation of the works and power of God." Moreover, nature's "majesty and grandeur, charm and grace, strength and omnipotence" and even immortality reside as well within the individual, fitting the person to worship the Creator.[27]

Ralph Waldo Emerson

In keeping with the discussions of the place of landscape and of natural history in our psyche, Emerson observed that our view of nature is all important in its determinative effect on society itself. As he puts it, reason on its own may lead to science, but *feeling* takes us to a true understanding of the real world. Indeed, every individual must have a role in determining his or her destiny. Emerson was a leader of American transcendentalists, "who invoked nature as a norm by which to judge society," with nature validating "the insights of the individual who took the natural world seriously."[28]

Emerson is among those who addressed the matter of the natural in counterpoint to the urban. His own "moment of insight" occurred "not in a rural retreat, but while he [was crossing] a town common at twilight." He, for one, recognized our "deep uneasiness" in face of the actuality of "a latent fear" of an "inherent difference between the natural and the urban." In response, Emerson's philosophy for his nation as well as himself was "not to choose between literature and life, abstract and concrete, city and country, but to discover a way to unite rural and urban for spiritual and social advancement." While the countryside afforded solitude "as a path of self-awareness," without (urbane) society "we shall feel a certain bareness and poverty."[29]

Thus is set out in the simplest of terms the ambivalence we all experience. Emerson was writing to a primarily urban audience, but he was not advocating literal withdrawal. Instead, he was calling upon his readers "to reach down and discover the organic principles of their own being." To Emerson city and nature were inextricably linked on a circle from which we can find the perspective affording insight into that organic relationship. Moreover, if our cities "seemed to be separating from nature," it is a reflection of our own disunion within the self. To "mend the spiritual rift" the reformer must become a mediator between the two worlds and point us "to a true understanding of the relation among the self, the products of [our] action, and the universal pattern by helping to sharpen the inward and outward eyes."[30]

Henry David Thoreau

Thoreau has been described as both pastoralist and lover of the wilderness and raw nature. But he, too, subscribed to Manifest Destiny and to instilling human strength and naturalism through "redeeming the meadow." The ubiquitous ambivalence is very much present: Thoreau worshipped nature, but as a transcendentalist, he would also seek to overcome nature in the name of the spiritual. As Thoreau himself put it in *Walden*: "We are conscious of an animal in us, which awakens in proportion as our higher nature slumbers."[31] Whatever the proper balance between animal and some human "higher nature," Thoreau found "the essence of freedom" resident not in culture but in nature. The closer we "live to nature, the more likely [we] are to realize [our] freedom."[32]

Just a sampling of his writings reveals the Simple Life in three contexts of interest here: through the pastoral and atavistic and in terms of place. All lead him with his reader to connection through a sense of the aesthetic. In the process he intermingles the observations of the naturalist with a more poetic appreciation phasing with maturity into a mysticism identified with the transcendental.[33]

In *Walden* Thoreau proceeds from the pastoral to connections as he speaks of mentally "buying" local farms. As a result of this form of purchase, the buyer never loses the subject landscape. At the same time, Thoreau reveals a jaundiced view of the realities of farming, not always included with the rhapsodies over the Simple Life. The poet, he observes, frequently withdraws, "having enjoyed the most valuable part of a farm," leaving the "crusty owner" supposing he had taken no more than a few wild apples. The poet's "rhyme" is "the most admirable kind of invisible fence" about the farm: impounding, milking, skimming it. The poet "got all the cream, and left the farmer only the skimmed milk."[34]

Elsewhere, Thoreau thoroughly mixes the pastoral and atavistic with a touch of the Wild Man, in the person of Amerindians or as outlaw, with his message for civilization. Thoreau begins with the "innocent pleasures of country life," specifically that of making "the earth yield her increase, and gather the fruits in their season." But, he asserts, "the heroic spirit will not fail to dream of remoter retirements and more rugged paths." We would "sometimes ride the horse wild and chase the buffalo." And the Amerindian Wild Man's "intercourse with Nature is at least such as admits of the greatest independence of each."[35]

Thoreau notes that while gardening has its pleasures, it lacks "the vigor and freedom of the forest and the outlaw. There may be an excess of cultivation as well as anything else, until civilization becomes pathetic."[36]

Almost in the same breath, with reference to the prospect of "civilizing" the Amerindian, Thoreau observes it would be no improvement and, moreover, identifies the Indian as something of a Wild Man: "By the wary independence and aloofness of his dim forest life he preserves his intercourse with his native gods, and is admitted from time to time to a rare and peculiar society with Nature." Indeed, a London or Boston is "refreshed by the mere tradition" of the "more primeval savages'" wild fruits.[37]

With regard to the more purely atavistic, a contemporary commentator observes Thoreau observing his own kind. The ambivalence is evident:

To eat a woodchuck raw is to behave as a savage, even perhaps as an animal living unconsciously and spontaneously, and instinctively following primordial patterns. In the woods there is no polite conversation over glasses of wine and veal served on fine china . . . , but a . . . realization of the vital center — nature as will — of organic life. Thoreau is the most civilized of men, and he knew it. Although he stops short of adopting the woodchuck as a totemic symbol, he . . . verges on recovery of the Paleolithic mind.[38]

In Thoreau's time others were already entering the wilderness more for recreation than for sustenance, coming "to sketch or sing," moving Thoreau to assert that "our life should be lived as tenderly and daintily as one would pluck a flower." He found it strange, however, "that so few ever come to the woods to see how the pine lives and grows and spires, lifting its evergreen arms to the light, to see its perfect success "instead of lumber, boards, and houses. Returning to poetry, Thoreau claims the poet makes the truest use of the pine and that he himself sympathizes with "the living spirit of the tree."[39]

Any number of additional writers of transcendental prose and poetry could be included here, but the important impressions have been conveyed. Before moving on, however, a contemporary description of the works of Edgar Allan Poe and Nathaniel Hawthorne is revealing. Both are described as "masters of the impressionistic sketch," with particular reference to their use of chiaroscuro, the blending of light, shade, and shadow. Most affective are forest light, light and dusk, moon and firelight. In mastering chiaroscuro, the literary or visual artist "fuses the physical and psychic dimensions of landscape," evoking a mental state receptive to "supernatural or preternatural dimension of reality."[40]

Whether in reality or through the imagination of others, along with John Burroughs, we "go to nature to be soothed and healed, and to have [our] senses put in tune once more."[41]

NOTES

1. Machor, *Urban Ideals and the Symbolic Landscape of America* (1987), 81.

2. Blum, *Picturing Nature* (1993), 15, 17.

3. Conron, *The American Landscape* (1974), xvii–xviii.

4. Ibid., 143.

5. For a mere sampling of "coffee table" volumes celebrating landscapes as special places see Paul A. Rossi and David C. Hunt, *The Art of the Old West*, 1971, New York: Alfred A. Knopf, Inc.; Ansel Adams, with Mary Street Alinder, *Ansel Adams: An Autobiography*, 1985, Boston: Little, Brown and Company; A. Aubrey Bodine, *Chesapeake Bay and Tidewater*, 1954, Baltimore: Hastings House; Elise Maclay, *The Art of Bev Doolittle*, 1990, Trumbull, Conn.: The Greenwich Workshop, Inc.; Elise Maclay, *Bev Doolittle: New Magic*, 1995, New York: Bantam Books; Tom Slayton, *Sabra Field: The Art of Place*, 1993, Shelburne, Vt.: Chapters Publishing, Ltd. and Montpelier, Vt.: *Vermont Life Magazine*; Wanda M. Corn, *The Art of Andrew Wyeth*, 1973, Greenwich, Conn.: New York Graphic Society Ltd.

6. Conron, *The American Landscape*, xx.

7. Michael Drayton, in ibid, 108–109.

8. Schama, *Landscape and Memory* (1995), 189–190.

9. Conron, *The American Landscape*, 149.

10. Ferkiss, *Nature, Technology, and Society* (1993), 77.

11. Blum, *Picturing Nature*, 7.

12. Ibid., 24, 54.

13. Ibid., 55, 57, 59.

14. Ibid., 227, 210–211, 237.

15. Ibid., 60, 62, 63, 66, 88, 115.

16. Marsh, quoted by Lowenthal in his introduction to Marsh (1864, 1965).

17. Orr, *Ecological Literacy* (1992), 125, 126.

18. Fairchild, *The Noble Savage* (1928), 1.

19. Holbrook, preface to Jamil, *Transcendentalism in English Romantic Poetry* (1989), x.

20. Fairchild, *The Noble Savage*, 57.

21. From Warton's "The Enthusiast or The Lover of Nature," in ibid., 60, 61.

22. Jamil, *Transcendentalism*, 88, 92, 106.

23. Ibid., 22, 23.

24. Ibid., 128.

25. Ibid., 128, 130, 109.

26. Ibid., 50, 52.

27. Ibid., 195, 196, 182.

28. Ferkiss, *Nature, Technology, and Society*, 74–75.

29. Machor, *Urban Ideals*, 156–157, 158.

30. Ibid., 160, 161.

31. Ferkiss, *Nature, Technology, and Society*, 75–76.

32. Oelschlaeger, *The Idea of Wilderness* (1991), 165.

33. See, e.g., Zeveloff et al., *Wilderness Tapestry* (1992), 42–43; Oelschlaeger, *The Idea of Wilderness*, 161, 165, 169.

34. Thoreau, *A Week on the Concord and Merrimack River — Walden; or, Life in the Woods — The Maine Woods — Cape Cod* (1985), 388.

35. Ibid., 46.

36. Ibid., 45–46.

37. Ibid., 46–47.

38. Oelschlaeger, *The Idea of Wilderness*, 161.

39. Thoreau, *The Maine Woods*, 685.

40. Conron, *The American Landscape*, 272.

41. Mattingly, *Songs of the Earth* (1995), unpaginated.

7

Heroes, Wild Men, and the Noble Savage

THE PLACE OF LEGEND AND LORE

We may be unique in our insistence upon clear distinctions among history, myth, legend, and folklore. As a "part of the inherited conglomerate of accepted beliefs, values and attributes" giving a culture its identity, legends provide valuable evidence of their society as well as insights into human nature. Unlike myth's traditions combining nature, destiny, gods, and humankind, legend focuses almost entirely on individuals, albeit ones who are larger than life. Like myth, legend may serve moral or exemplary purposes. Early western legend associates persona with special places — real, mysterious, hidden, or lost.[1] The starting point of this continuing theme is found in Carl Jung's collective unconscious and archetypes.

The Collective Unconscious and Archetypes

Rudiments of Jungian Psychology

Prophetically for environmental stewardship, Jung speaks of psychology, medicine, and natural science being estranged from our "philosophy, history and classical learning," leading to a "distance between nature and mind" increased by arcane jargon. He studied "primitive psychology, mythology, archaeology and comparative religion" for their "priceless analogies," specifically for their associations in his patients' dreams.[2]

Jung describes our many integrated parts, notably including the psychic as well as physical. To this pioneering psychologist "[t]he continuity of nature knows nothing of those antithetical distinctions which the . . . intellect is forced to set up" in aid of comprehension. The distinction between mind and body is also an artificial dichotomy, more of the intellect than of nature. Jung cuts through the jargon to note that "it is only the philosopher who does not know" what thinking is; "no layman will find it incomprehensible." However difficult it is for science to "define such notions . . . "they are easily intelligible in current speech."[3]

But Jung also warns that our psychic life, "the dubious gift of civilization," is full of problems; our mental images are "almost completely foreign to the unconscious, instinctive mind of primitive man." Consciousness may only deal with culture, but "instinct is nature and seeks to perpetuate nature." In short, ambivalence is inherent to our nature; we are constantly seeking a workable balance between the wild and the civilized. To Jung attempts to explain the psychic in terms of mere physical factors are "doomed to failure."[4]

The Jungian collective unconscious emerges in ubiquitous psychic activity often expressed in pictorial symbols springing from and satisfying a basic need. Through these symbols we reach back into our primitive past and reconcile it with contemporary consciousness. In Jung's vision of the collective unconscious, the "vast outer realm" is mirrored by "an equally vast inner realm." We stand between the two, facing first one and then the other.[5]

Jung defines the collective unconscious as "a certain psychic disposition shaped by the forces of heredity" and revealed in "eclipses of consciousness," including narcosis and certain forms of insanity as well as dreams. The collective unconscious is manifested as "[m]ythological themes clothed in modern dress." Significantly, Jung asserts that an artist's work reveals something of the contemporaneous collective unconscious, which "becomes a living experience" bearing upon our conscious outlook and important to all. It is "a message to generations" with the artist a "collective man" who "carries and shapes the unconscious, psychic life of mankind." The collective unconscious represents an immense inherited fund of factors finding expression in virtually countless "subliminal perceptions." It can also be a personification made up of ubiquitous, timeless identities known as archetypes.[6]

Selected Archetypes

The many archetypes are "buried and dormant" in our unconscious, waiting to be "awakened whenever the times are out of joint and a human society is committed to serious error." In the artist's work they meet the spiritual needs of the artist's society.[7] It is in the nature of archetypes to leave evidence of their presence as expressed in our "stories, dreams, and ideas" where "they become a universal theme, a set of instructions, dwelling who knows where, but crossing time and space to enwisen each new generation."[8]

Among those archetypes relevant here are the Anima (the Feminine), the Green Man, the Trickster, the Hero, the Redeemer (Savior), the Wild Man and Wild Woman, and the (Internal) Lover. As will be suggested here and throughout the text, any and all of these entities can occur separately or together in one individual of either sex.

Terry Tempest Williams defines the Feminine as a reconnection to self and a commitment to internal wildness, including instincts, creativity, and destructiveness — "our hunger for connection as well as sovereignty, interdependence and independence." This archetype teaches us to find "our way back home" through experience, "the psychic bridge that spans rational and intuitive waters." The Feminine preserves "mystery, and mystery inspires belief."[9]

Ordinarily a composite of a man's head and vegetation, the Green Man, guardian and revealer of mysteries, signifies irrepressible life, renewal, rebirth. According to William Anderson, the Green Man's contemporary reappearance in art and as a symbol of the environmental movement is profoundly significant. This power-laden image, after a long absence, is needed because he "knows and utters the secret laws of Nature." This archetype may have been introduced to us by Dionysus and may be found in Saint George, Robin Hood, the Trickster, and the Wild Man. Significantly, Robin Hood's appeal includes his dangerous but carefree life of atavism.[10] In addition to being the Green Man, Robin Hood and Dionysus are both the Wild Man and the former is also the Hero and the Trickster, revealing again how archetypes interconnect with each other in the same persona.

It is the Hero who, on our behalf, accepts deprivations, sacrifices comfort, "throws off status or wealth, all in an effort to secure the safety and well-being" and empowerment of others.[11] Heroes are clearly ubiquitous and are said to share a constellation of features. In the fifties, twenty-two characteristics of the "classical Hero" were listed. While no single hero may fit the entire pattern, some combination of many of the characteristics is to be expected.

The Hero's mother is a virgin of royal blood and his father a king, often closely related to the Hero's mother. The conception is somehow unusual, with the Hero reputed to be the son of a god. At birth there is an attempt on the Hero's life on the part of the father or maternal grandfather, but the Hero is spirited away to be reared by foster parents in a far country. Nothing much is told of his childhood, but upon entering manhood the Hero returns or proceeds to his future kingdom. After attaining victory over the present king or some extraordinary being, the Hero marries a princess and assumes the throne. For a time he reigns uneventfully and prescribes laws, but eventually he loses favor with the gods or with his own subjects. He is driven from the throne and his city only to meet a mysterious death, often on a hilltop. Although he is not buried, one or more holy sepulchres are established in his name. Heroes are often abnormally slow in aging, even when parents of adult offspring. Stock characters around Heroes retain the age at which they are introduced into the epic.[12]

Interestingly enough, heroes may generally be subject to antipathy by reason of their heroic nature, which does not necessarily include good citizenship. The Hero is an outcast from settled society, because he is a disturbing influence that must be suppressed for society to maintain order. "Free spirits, high-hearted, bold and daring," are not wanted in an orderly world,[13] witness the cinematic version of John Rambo, especially in the second entry.[14]

Hero and Wild Man are a frequent combination. The Wild Man has communications with animals, each one of which opens a new possibility, and one cannot be a Hero without being in touch with either an animal nature or the Wild Man within. Significantly, however, once in touch with the Wild Man and having received his gift, mythic Heroes and Heroines never keep that gift; they return it or give it away.[15]

HEROES, WILD MEN, AND THE NOBLE SAVAGE

Three persona are particularly relevant to our relationship to the natural world. The Hero archetype represents that which we hold brave and noble; the Wild Man archetype is a popular image of atavism; and the noble savage, often an amalgam of the Hero and the Wild Man, is more an image in a culture's contemporaneous popular arts than an archetype. Frequently, these three images are merged, although not all wild men rise to the level of the corresponding archetype.

The Hero and the Simple Life

Any number of heroes from European folklore find themselves torn between two worlds, which can be metaphors for civilization and some form of a Simple Life. Representative among them is the Norse tale of Helgi Thoreson, son of a chieftain, who became separated from his party. Upon seeking shelter in a pine forest Helgi entered a place of mist where he encountered a mystical party of women. Finding them enchanting, Helgi accepted their hospitality to remain with the leader. The other women vanished and Helgi remained subject to the enchantment of woman and place but never forgot his home. Once he broke the enchantment to return home, Helgi commenced to grieve for the woman. Finally, he vanished. Presumably resident in her land of mist, he returned on occasion to his own world in response to a "never-ending call across the border of worlds."[16]

There is more than one lesson here. Among them is the division of self when venturing between worlds. To cross the boundary is to become a ghost who belongs in neither one place nor the other. Whichever is chosen, there is a longing for the other,[17] an experience that apparently transcends both time and space. According to our cultural anthropologists, all such stories are a mirror for our own culture. The stories reflect and play with cultural realities. Through the stories, such as the hero outside the tribe, we learn a people's perspective on themselves — their "institutions, values, beliefs, and assumptions."[18]

There is in fact a genre of "Fenian" heroes, who anticipate Tarzan of the Apes and others in being "genuinely part animal." The animal portion, inherited from parent or foster-parent, is an expression of the ambivalence of the outlaw and of ourselves. Such persona are "ominously alienated" but "constantly exerting pressures" on society.[19]

Ireland's Finn is a prestigious poet, a proclaimer of truth, custodian of tradition, and possessor of otherworldly knowledge. He is also an outcast, caught up in a pattern of opposition and reconciliation. The poetry associated with his persona, of the *fénnidi*, is of the wilderness, not merely expressing a love for nature but a strong preference for wilderness and the savage life in contrast to the comforts of society. Wilderness holds a special source of knowledge not present in the civilized world.[20]

Such heroes are closely related to wild men and noble savages. Significantly, Finn, and others, are never to be content: "Because he can penetrate worlds and form relationships so easily, the anomalous Finn cannot acquire true and lasting membership in either human or supernatural society."[21]

The Wild Man

The notion of the Wild Man dates back to biblical times and ideas of wilderness, defined as areas not domesticated — desert, jungle, mountains. Interestingly enough, Wild Man's identification with a specific place dissipated as real wildernesses were penetrated and came under human management. Today he has been brought out of the distant wilderness and placed within us, where he is "lurking" and "clamoring for release."[22] Significantly, the Wild Man is far from unique to Western civilization. His counterparts are to be discovered throughout the world. Whatever his form, the wild man, if not the archetype itself, represents the freedom from responsibility that we all desire to some extent. Most of us cannot indulge in that freedom and must enjoy it vicariously. These images carry us into an atavistic relationship with nature more deeply than we are likely to venture. Hence, their fascinating and enduring call to us.

From the Middle Ages on, the ubiquitous wild man has been an escape from "civilized evils."[23] One is Shakespeare's Caliban, described as a combination of islander cannibals, Europe's Wild Man, the exploited savage, and the subjected peasant of Europe. He was actually something of a mirror image of Shakespeare's contemporaries in civilization.[24]

Cervantes, in inventing the novel, is also a progenitor of science fiction, and his *Man of Glass* is a kind of Wild Man. Having drunk a poisonous love potion Cervantes' protagonist suffered the delusion of being fabricated of glass. His extreme fragility led him to comprehend the world in a manner uniquely different from others. Seeing through all human contrivances, the protagonist "uttered penetrating witticisms on the charlatanry" all about him. Upon being healed of his delusion, the restored man sought to preach, but there was no longer any interest in what he had to offer[25] — he was, after all, no longer a Wild Man.

A feral child of the early eighteenth century inspired an outpouring in a variety of literary genres, including a satiric book by Daniel DeFoe and a poem entitled *The Savage*. This poem called upon the elite courtiers of the time to receive this "[y]outh unform'd, untaught," "[w]ild, and a [s]tranger to his kind." The courtiers were exhorted to teach him reasoning and manners and to civilize him. But dire warnings follow on the heels of exhortation: Do not teach him to beguile, to ensnare, to turn him to vice, envy, pride, or avirice. Do not, in short, taint him with civilization's wiles. Allow him his innocence and freedom; remember just *who* is the savage.[26]

The Noble Savage

At some point — the seventeenth century has been suggested — the noble savage emerged. He is distinguished from the wild man by uncorrupted virtue and innate nobility. His characteristics are extracted from whatever primal people contemporaneous explorers were encountering. A cult, with political implications, of the noble savage had arisen by the eighteenth century. Its tenets combined a "belief in the natural goodness of man" with criticism of "European cultural assumptions." Rousseau's noble savage was no "stock" character. He was the internal savage, one whose presence was at once necessary and good. Most notable among noble savages of this era are Amerindians.[27]

That the noble savage and his kin are potent and important images is revealed through the array of scholars who have turned to them for explanations of human nature. A long-accepted practice casts these images as devices for social satire and political commentary. The noble savage is the "antithesis of culture and civilization, and therefore for whatever we have lost or repressed in becoming civilized."[28] He has been defined as "any free and wild being who draws directly from nature virtues which raise doubts as to the value of civilization." He protests "the evil incidental to human progress" while looking "yearningly back from the corruptions of civilization to an imaginary primeval innocence."[29]

Over time, the noble savage remains essentially the same in his courage, loyalty, stoicism, aloofness, and tendency to compare his culture (favorably) with the European, tending to become the Wild Man. The character of the noble savage is determined by "changing cultural attitudes toward the natural and the primitive which he symbolizes." His is a nostalgic image representing a lost Eden, a lost innocence, a mythical past and a longed-for escape. He is the suppressed internal Dionysian savage.[30]

The amalgam of hero, wild man, and noble savage can be found in early novels such as *Robinson Crusoe, Swiss Family Robinson,* and Chateaubriand's *Atala* and *Rene. Atala* tells of a Frenchman, Rene, who had been adopted by the Indian, Chacta. Each of the three principal characters is torn between wilderness and civilization. Chacta observes in his adopted son "the civilized man who has become a savage; you see in me the savage man whom the Great Spirit . . . has wished to civilize." From opposite extremes, "you have come to rest in my place and I have sat in yours; thus our outlooks should be totally different. Who, you or I, have gained or lost the most by this change of station?"[31]

While well treated by two Spaniards after being mistreated by their countrymen, Chacta eventually developed a distaste for the city and commenced to pine: "Sometimes for hours I remained motionless, gazing at the summit of the faraway mountains; sometimes they would find me on the bank of a river, sadly watching it flow. I would picture to myself the woods through which those waters had passed, and my soul would be entirely steeped in solitude."[32]

The ambivalence that attends the tension between civilization and wilderness and other forms of the Simple Life emerges in an unlikely place: Mary Shelley's original *Frankenstein.* The movingly pathetic conflict endured by the "monster" is more implied than expressed, but Boris Karloff succeeded in capturing much of it in the American films of the thirties, *Frankenstein* and, more effectively, *Bride of Frankenstein,* which takes more from Mary Shelley's book than does its predecessor.[33]

THE RETURN OF THE WILD MAN

In more recent times Tobias Schneebaum recalls being haunted by this archetype upon attending an exhibit on the Wild Man of Borneo. Schneebaum reflects on the constant presence of the archetype within himself: "I think I'd seen a loneliness in him," he was different as I was, "never knowing why or in what way." Schneebaum experienced the archetype in dreams and daydreams and may speak for us all: "I run, I run, and the forest is all around me. It nurtures me and fills me with longing for that elemental apparition. . . . I see him in an isolated time, living fierce and strong, his body hewn of ironwood, my own ancestor, my own self, flying through the trees. And when he moved within me and shook me, when he flows along my bloodstream, I ache and yearn to yell out, . . . to release the voices, the forces, . . . the destinies, the terrors and beauties that make up the meaning I give myself."[34]

A less intense description is that of a post–World War II Lithuanian forest guide, who had suffered through the war's occupations and withdrawals in Poland. His "startling blue eyes smiled from a face the color of tree bark." Throughout the ordeal this man clung to memories of his nation's forests and expressed no care for states — "This is my State — he smiled, waving airily at the trees — nature; you understand: the state of nature."[35]

If we were truly born to this connection, then turning away from it has to be wrenching and the maintenance of the distance a painful rejection of identity. It is the collective unconscious that shelters our "compacted ecological intelligence." The two parts of us, wild and civilized, must be reunited, not only for the sake of sanity but also to allow "greater evolutionary adventures."[36] The fictional wild man's need to reconcile his wildness with his civilization is a metaphor for that human need.

The Reality of Feral Children

True feral individuals are extremely rare in the real world. A very few individuals, usually children, have been reported to have been isolated from their culture, often to be nurtured by a domestic or wild species. These are the human equivalent of feral animals. In reality, it would appear there is none who would ever have been mistaken for a noble savage.

Feral, as applied to human individuals, became a part of organized science by name and concept in 1758 when Linnaeus cited eight cases.[37] The term is believed to have originated in France in 1750, and the concept had already become heavily romanticized as early as 1754. By the end of the eighteenth century a total of fourteen cases had been recorded.[38]

While feral children do not compare well with the typical fantasy, they do reveal some interesting parallels, suggesting some veracity in genre entries and in terms of our fascination with the concept. Some psychologists and anthropologists hold that we must grow up among other humans to become human. Otherwise we are totally isolated or take on the characteristics of those with whom we do mature from early childhood. It appears that too long an isolation from humanity or too strong an animal conditioning will permanently inhibit human development, and, in the latter instance, traces of the animal conditioning are never completely lost.[39]

In 1754 Rousseau recounted five cases of feral individuals. All of them were quadripedal rather than bipedal. One, found in 1344, was reared by wolves. Another was reared by bears and showed neither sign of reasoning ability nor power of speech.[40] By 1940 there were thirty-one reported cases of feral individuals and by 1964, fifty-three. An amazing array of foster animals is listed: wolves, bears, baboons, panthers, gazelles, snow-hens, and, yes, apes. Among domestic species are pigs, sheep, and cattle.[41]

Almost all the reported feral individuals were speechless, although many uttered animal sounds. They continue to be reported as quadripedal. They are reported to possess extraordinarily keen senses. Hearing and the sense of smell are highly developed; individuals are recognized by scent. By way of contrast, an almost complete insensitivity to temperature changes and extremes is reported. One such child came to describe his initial views from his window as ugly, perhaps because he was unable to distinguish individual objects.[42]

The human qualities of feral individuals tend to remain underdeveloped. The impression conveyed is one of complete selfishness. Emotions, however, are lacking. Neither anger nor impatience is displayed, and laughter and smiles are totally absent. Only one feral individual reportedly showed concern for future needs. Very few learn to like human society, several reportedly persisting in a longing for the wild. Clearly, any need for companionship takes the form of love for animals, with a preference for the foster-species or one closely related. It is fascinating to learn that wild animals lose their fear of humans in the presence of feral children.[43]

Victor of Aveyron

Thanks to his mentor, Jean Itard, a teacher of deaf-mutes, Victor's is the most fully reported of the early cases. His age was estimated at twelve when he returned to civilization. He was thought to have been abandoned at four or five. He was covered with scars, some of them animal bites. He was described as melancholy, taking a certain joy from the "violence" of nature. Itard was never able to teach him speech.[44]

Itard's descriptions of Victor's responses to the world around him are simultaneously fascinating and moving and relevant here: After dinner he would sip at a glass of water while "standing at the window, eyes turned towards the countryside as if in this moment of sheer delight this child of nature seeks to unite the only two things which remain from his lost freedom, a drink of clear water and the sight of the sun on the countryside."[45]

One cannot help but wonder with Itard what might have been passing through Victor's mind when, at the end of the warm seasons the boy entered the garden where he would walk about before sitting at the edge of water where "all his convulsive motions, and the continual rocking of his whole body diminished, . . . to give place to a more tranquil attitude; and how insensibly his face . . . took the well-defined character of sorrow, or melancholy reverie, in proportion as his eyes were steadily fixed on the surface of the water, and when he threw into it, from time to time, some remains of withered leaves." And on "a moon-light night," when the light entered his room, Victor would wake and go to the window. "There he remained, during a part of the night, standing motionless, his neck extended, his eyes fixed towards the country illuminated by the moon, and, carried away in a sort of contemplative [ecstacy], the silence of which was interrupted only by deep-drawn inspirations, after considerable intervals, and which were always accompanied with a feeble and plaintive sound."[46]

That Victor not only missed his wild life, but also longed to be human is apparent in Itard's chronicle.[47] The ultimate impression Itard conveys is one of pathetic mutual effort resulting in the taming of the wild child without civilizing him. Thus, Victor lost his wild world without ever being able to enter the civilized one. Could anything be more devastating?

Amala and Kamala

These apparent sisters are recently reported and photographed in India.[48] When they were found, Amala's age was estimated at three and Kamala's at five or six. It was postulated that they were taken by a she-wolf who had lost her own litter. This female had been approaching humans persistently and was leaving marks of blood-stained milk. The girls were described as wild in appearance with ferocious behavior. In repose their faces were completely blank. They were always more comfortable on all fours than standing or walking erect.

The children were noisy sleepers who slept lightly and awoke to the least sound. Any hint of danger brought them to the alert. Just before dawn they would become restless for the open. At night they would howl with a cry described as neither animal nor human but presumed to be a call to the wolves to advise the animals of their location. When irritated the girls uttered a peculiar, high-pitched growling.[49]

The morose, apathetic behavior of Amala and Kamala was attributed to complete disorientation among humankind and to a desperate loneliness interpreted as a longing for their former life and the companionship and protection of the wolves. Amala soon died of nephritis, but Kamala had been separated from the younger girl and survived. At the loss of her companion, Kamala remained expressionless but did persist in a peculiar, repetitive cry and was particularly morose and unapproachable.[50]

Thereafter Kamala became something of a normal child. She commenced to understand and to use language, although she could comprehend far more than she could express. For a time she made rapid progress. By the time she was eleven or twelve she had a vocabulary of some thirty words. She preferred sign language and rarely initiated speech, which she used only sparingly. With time her face showed animation but only when she was speaking. The girl could even be sent on simple errands. For a while she could be heard singing or chanting. It became a habit and then ceased.[51]

Kamala eventually dissociated from the dogs and they from her. To a limited extent she began to identify with her fellow orphans.

Unfortunately she became ill and died in two years.[52] With her died the opportunity to communicate with the unique adult who must emerge from the background of feral childhood.

The Literary Lineage

In literature, birds are a symbol of freedom; spontaneous, they live for the moment. We, too, long for a release from worries and difficulties and rationality. We dream of freedom. The literary lineage of wild or feral humans, who represent that freedom, is lengthy and broadly cast. It appears we are willing to express our bestial natures, to become savages in the process.[53] Estés tells of the feral woman,

One who was once in a natural psychic state — that is, in her rightful wild mind — then later was captured by whatever turn of events, thereby becoming overly domesticated and deadened in proper instincts. When she had opportunity to return to her original wildish nature, she too easily steps into all manner of traps and poisons. Because her cycles and protective systems have been tampered with she is at risk in what used to be her natural wild state. No longer wary and alert, she easily becomes prey.[54]

There is an Amerindian tale of Wolf Woman who long ago hurt her leg and was unable to travel with her tribe. With the approach of winter they left her beside a stream with dried meat and a horse, but the horse ran off and the woman soon ate her food. When a wolf repeatedly came near her camp the woman eventually explored her den and was accepted by the wolf to the point of being fed and eventually joining the hunt. The woman remained among the wolves for several winters and was accepted as a relative by the grown cubs. She spoke their language. When her tribe returned, she remained aloof for a time. When she again joined them, she "had the ability to tell the band of approaching blizzards and storms, for she could interpret the songs of the wolves singing to the moon."[55]

Roszak urges us to reconnect with both our biological and our ecological selves and notes that at least one Freudian psychiatrist holds the belief that a child's mind "recapitulates the psychic phylogeny of life on Earth" before becoming the integral, independent entity that is the mature person. At least one Romantic poet suggests "that the acquisition of speech marks the end of our animistic sensibility"; thereafter "the child grows away from inherent intimacy with nature."[56]

NOTES

1. Cavendish, *Legends of the World* (1982, 1989), 9–13.
2. Jung, *Modern Man in Search of a Soul* (1933), 30, 65.
3. Ibid., 30, 65, 74, 89.
4. Ibid., 222, 242.
5. Ibid., 71–72, 120, 126.
6. Ibid., 165, 169, 186, 241.
7. Ibid., 171.
8. Estés, *Women Who Run with the Wolves* (1992), 327.
9. Williams, *An Unspoken Hunger* (1994), 53.
10. Anderson, *Green Man* (1990), 33, 14, 28, 34.
11. Swimme and Berry, *The Universe Story* (1992), 59.
12. Raglan, *The Hero* (1956, 1975), Chap. XVI.
13. Slate, "Edgar Rice Burroughs and the Heroic Epic" (1968), 122.
14. *First Blood* (1982), Ted Kotcheff, director; *Rambo: First Blood Part II* (1985), George P. Cosmatos, director.
15. From notes taken at a seminar on the Wild Man and the Wild Woman coconducted by Robert Bly at Chicago's C. G. Jung Center in April 1983.
16. Anonymous, *The Enchanted World — Fabled Lands* (1986), 99–104.
17. Ibid., 104.
18. Nagy, *The Wisdom of the Outlaw* (1985), 13.
19. Ibid., 44.
20. Ibid., 19, 28, 36.
21. Ibid., 121.
22. White, *"It's Your Misfortune and None of My Own"* (1972), 7.
23. Shepard, *Man in the Landscape*, 174.
24. Miner, "The Wild Man Through the Looking Glass," in Dudley and Novak (1972), 96–97.
25. Boorstin, *The Creators* (1992), 302–303.
26. From Miner's quotation of the poem, in Fairchild, *The Noble Savage* (1928), 186.
27. Symcox, "The Wild Man's Return," in Dudley and Novak (1972), 228, 229, 233–234.
28. Ibid., 281.
29. Fairchild, *The Noble Savage*, 2.
30. Symcox, "The Wild Man's Return," 282, 287.
31. Chateaubriand, *Atala*, 20.
32. Ibid., 21.
33. *Frankenstein* (1931), James Whale, director; *Bride of Frankenstein* (1935), James Whale, director.
34. Schneebaum, *Wild Man* (1979), 1–3.
35. Schama, *Landscape and Memory* (1995), 61–74.
36. Roszak, *The Voice of the Earth* (1992), 304, 305.

37. Zingg, "Feral Man and Extreme Cases of Isolation," *American Journal of Psychology* 53 (1940), 39.

38. Malson, *Wolf Children and the Problem of Human Nature* (1972), 40–42.

39. Zingg, "Feral Man and Extreme Cases of Isolation," 514; Malson, ibid., 9.

40. Malson, ibid., 38–40.

41. Zingg, "Feral Man and Extreme Cases of Isolation,"11; Malson, ibid., 38–40.

42. Malson, ibid., 48.

43. Ibid., 36–44, *passim.*

44. Itard "The Wild Boy of Aveyron" (1799), translated in Malson, ibid., 91.

45. Malson, ibid., 151.

46. Ibid., 104.

47. Ibid., 171–172.

48. Maclean, *The Wolf Children* (1977).

49. Ibid., 87–88, 161, 167.

50. Ibid., 135–136.

51. Ibid., 150, 163, 166–170, 172, 199–201, 205.

52. Ibid., 125, 213–219.

53. Hermunsgård, *Child of the Earth* (1989), 90, 68–69.

54. Estés, *Women Who Run with the Wolves*, 214.

55. McGaa, *Mother Earth Spirituality* (1990), 164–165.

56. Roszak, *The Voice of the Earth*, 295–296, 298.

III

REFLECTIONS OF THE NATURAL WORLD: WHAT WE LEFT BEHIND

8

Lessons from beyond America's Western Tradition

All human cultures have their "own humus of custom, myth, and lore." Like the ancient forests, they have their depths and densities.[1] Beyond practicality, survival depends upon a "rapport with the all-pervasive spirit powers manifested throughout the phenomenal world," often paralleled in the symbolic rituals in which peoples engage.[2] Today we have lost a kind of transparency, the ability to see through the world to perceive "greater realities" moving "behind and within" us and "seen in this and that, here and there as if through a lens."[3]

Across cultures, there is a perception of nature as "a harmonious universe" that is "grand and spectacular, fragile yet eternal," replacing what was "a dark and ominous force to be avoided or propitiated" or a "sacred but still dangerous presence to be approached with circumspection."[4]

Hunting peoples, in particular, share a religion centered on "the notion that a spiritual landscape exists within the physical" one. On occasion "one sees something fleeting in the land, a moment when line, color, and movement intensify and something sacred is revealed," and is led to believe "in another realm of reality" corresponding to our own, yet different. It is our profound loss that science has been allowed to overshadow such "esoteric insights and speculations," a poetic regard of land with its coherence, transcendence, meaning, and power.[5]

Any religious commitment to a unique "founder-sage" leads to an awareness that any borrowing from others is no more than apparent. If such founders are expressing some aspect of a single ultimate reality, it should come as no surprise that their respective holy mysteries prove to share much in common.[6]

Pope John Paul II speaks of a spiritual history that includes all religions as a reflection of human unity with regard to our ultimate, eternal destiny. The Pope speaks of a community of all peoples who are of one origin as well as one destiny. He finds *semina Verbi* ("seeds of the Word") present in all religions and asserts that the Holy Spirit works effectively outside "the visible structure of the Church" while Christ (the Second Person of the Christian Trinity) "has His own ways of reaching each" people.[7]

If the Catholic Church is open to other versions of the Holy, we should not be reluctant to acknowledge and assimilate truth from earlier generations and alien peoples. Truth is the highest value among its seekers and it never "cheapens or debases" those who reach for it but "ennobles and honors" them.[8] "God, the omnipresent and omnipotent, is not limited to any one creed"; "wheresoever ye turn, there is the face of al-Lah."[9] Jalalal-Din Rumi, founder of Islam's Sufi order, asserts we are all searching for God and advises us to listen to how the reed "tells a tale complaining of separateness." He, too, bemoans his parting from the reed-bed and tells us that each "left far from his source wishes back the time when he was united to it."[10]

The poetry of Islam's Masnawi speaks of the agonies of separation and of finding a transcendental dimension and looking for hidden reality. Our ego blinds us to inner mysteries of all things, but once we get beyond ego "we are not isolated, separate beings but one with the Ground of all existence."[11]

Every culture possesses "an aesthetic moral aspect" and demonstrates "the power of the imagination to transcend group values" to reach out for the values of others and, through them, grow. As individuals we can surpass the norms of our own groups. From fear we can progress to awe and then to appreciation.[12]

It is said that the "attentive individual" can appreciate an array of cultures because we all share the "same biological senses and mind." Moreover, there are biological commonalities in how our "senses work together, and how the mind, in rather similar ways, builds on such . . . foundations to create symbolic spaces and complex cultural worlds" with their own commonalities.[13]

Fantasy is one source of enlightenment, of which Jung holds a high opinion, considering fantasy "the maternally creative side of the masculine spirit." Although many forms of fantasy may be held in low regard, all human works "have their origin in creative fantasy," deep and "closely bound up with the tap-root of human and animal instinct."[14] The quest for meaning is not solitary but demands an outward journey. It is through our communities that we can empathize with another's search for meaning as we suspend our frame of reference to confront others' spiritual, intellectual, and emotional worlds.[15]

PRIMAL CULTURES

There are always risks in looking from one culture into another. The expert observers will not agree among themselves and may all be wrong in any event. Obviously, as an individual and as a member of a "foreign" culture, each anthropologist will bring an array of biases and outright prejudices to the analysis of a culture. Misunderstandings of the culture itself and of its larger implications may result in distortions. More important to the subject culture, however, there can be practical repercussions.[16] Nevertheless, lessons can be gained without loss to those providing them, so long as care and respect are extended in the process.

Overview

Insights are available through the "early tribal period" shared by all humankind, when all of us "lived in a world dominated by psychic power symbols" guiding life "toward communion with our total human and transhuman environment." In those times we could sense a cosmic presence sustaining us beyond nature's surface reality. A sense of some "all-pervasive, numinous, or sacred power" could restore to our lives some profound security we have lost over the generations.[17]

In accepting connections between phenomena or between people and animals, Jung dismisses the word "mystical" because primal cultures see nothing mystical. It is perfectly natural, except to us because "we seem to know nothing about such psychic phenomena" that create a world in which we have a psychic as well as a physical presence.[18] George Perkins Marsh alludes to the multiple practical lessons discovered by "unschooled men" and the "teachings of simple experience . . . where natural philosophy [science] has scarcely yet spoken."[19] If we are to approach this genius of primal peoples in our quest "to hear and respond to the spontaneities of the universe," it is essential to establish such "a close relationship with the psychic depths of the universe."[20]

Estés tells of the seal woman and her child. In the mother's underwater home the child is educated in the ways of wild soul, representing a new order in the psyche. The child can bridge the two worlds but cannot remain in hers and must come back to earth to occupy a special role, somewhere between ego and soul. The "medial" woman or Wild Woman "stands between the world of consensual reality and the mystical unconscious and mediates between them," her own home being the midway point.[21]

Implications of Traditional Wisdom

To aid in efficacy conservation programs should draw upon traditional wisdom at three junctures: in determining which areas are to be protected, in the preparation and implementation of management plans, and in the sharing of income from tourism and related activities. Perceptions of wilderness are dependent upon a people's geographical, historical, and cultural circumstances affecting a hierarchy of needs. If a society is engaged in satisfying "lower needs," awareness of higher ones is inhibited, a lesson of import in the so-called Third World. Whatever contemporary circumstances may hold sway, historically "native cultures practiced . . . subsistence harvesting with respect . . . and in harmony with" wilderness, with their shelters blending into natural surroundings. Such peoples should be allowed to remain in their places where they are an effective component of the wilderness and a part of the natural balance. The law should accommodate this choice. No wilderness will survive without incorporation of its indigenous tradition, culture, and spirit.[22] A sampling of representative traditions follows.

African Peoples

Because human life is believed to have originated in Africa between a million and sixty thousand years ago, subsaharan Africa is an appropriate point of departure. It is worth noting that a special character is attributed to African art because of the small scale of the continent's societies and the immediate relationship between art and other social forms, with a wide variety of styles arising from ecological, historical, and social circumstances.[23]

One portion of the Serengeti ecosystem is the Mara of the Maasai, among the last nomads. The land is described by a visitor as "wild, interrupted country capable of capturing one's spirit like cool water in a calabash." Here the people and the land belong to each other—"the umbilical cord between man and earth has not been severed." Here Maasai cattle are pastured "next to leopard and lion." The people "know the songs of grasses and the script of snakes. They move like thin shadows across the savannah."[24]

In Benin of West Africa the world is rich with deities connecting the worlds of the living and the dead. In this form of immanence the trees, earth, and stones do not have gods, but deities are present everywhere. Numerous shrines of a variety of materials constitute a metaphor for connectedness. Vodun (Vo, to rest; Dun, to draw water) permeates life. Vodun's message is that in life we draw from a pool below, but it is proper to rest before doing so. The emphasis is on calmness and composure, the teaching to avoid rushing through life; to sit quietly and "accept the flow of events."[25]

That indigenous peoples are faced with many of the same problems besetting us is suggested by forest people of Madagascar. Known as "the robe of the ancestors" their forest represents something eternal that for centuries has provided water, shade, and medicines. Now that forest is in tatters, an ecological and economic tragedy in need of reversal for a variety of reasons at a variety of levels.[26] That the spiritual as well as the material realm is to be taken into account should not escape those who are taking any action.

The Malshegu of Ghana's Northern Region are an instructive case in point, with experience of some three centuries. Against the threats of the modern world the Malshegu are seeking the preservation of a small forest, home to the god Kpalevorgu. Thus far, the strength afforded by "traditional beliefs, embodied in the religious priest charged with protecting the abode of the god, have been sufficiently strong to prevent human interference in the forest." Open forest has been transformed into the partially-closed canopy forest, once a component of the original ecosystems.[27]

The Malshegu land includes the largest sacred grove in the region. For this grove the Malshegu have established and enforce a regime of land use rules and practices in the belief that failure to comply for the protection of the grove and to participate in festivals "will offend and dishonor" Kpalevorgu and risk misfortune for the offender, the offender's family, and even the community at large. Stories are told of others, including an American, who ignored warnings to violate the sanctity and improperly intrude upon certain protected areas only to become ill, suffer mental illness, or die.[28]

The regime is enforced through community vigilance under the direction of a religious leader with the support of the Malshegu chief and other leaders of the community and neighboring villages. The grove is respected even in face of local shortages. Success is attributed to the strong local religious beliefs, whereby isolating any forest activity from "traditional religious beliefs and practices" is difficult. In short, the Malshegu themselves are key to protection of the grove from nonbelievers. For us, the essential lesson here is that "[c]ommunity vigilance rather than formal or active policing is sufficient to enforce the regulations." Complementing this spiritual element are locally accepted protection guidelines and the regional importance of Kpalevorgu. Moreover, regional support is widening and strengthening the religious tradition, in turn strengthening local effort.[29] The individual must be engaged in the effort and interact with empathetic policy.

Australian Aborigines

Australia is home to a forty-thousand-year-old tradition of Aboriginal art figures and motifs. The Aboriginal art of oral storytelling goes back to the "dimmest times of the 'dreaming,'" presumably to Aboriginal settlement from Southest Asia.[30]

The Aboriginal way of life is that of hunter-gatherers. Tribes are of local groups based in territories each with its watering place, home to the group's settling ancestors and the place where all the "preexistent spirits" have been awaiting "incarnation and reincarnation." These founders and all their descendants are "forever kinfolk" notwithstanding separations in time, space, or custom. Ritual relationships are entered with species and the rain, reflecting recurrent food failures and drought. Each clan is related to a totem species. The lodges of the men are custodians of individual species and ancestral heroes with their mythology, sites, ritual, and symbols. Ritual reenactment of the creative past renders it operative in the present to assure life of the lodges' respective species and humankind. Aborigine myths and rituals constitute Dreaming, a "continuity of life unlimited in space and time."[31]

The walkabout is a pilgrimage to a succession of named places, each familiar from childhood and a part of the creation story. It is a journey into the interior, where the individual encounters his personal and tribal past simultaneously, thereby renewing contact. This is "a journey into time and space refreshing the meaning" of the individual.[32]

Australian architect Amos Rapoport mapped these mythological landscapes, identifying them as "unobservable realities" finding expression in "observable phenomena," whereby the land "makes the myth real." The stories are set in critically important local landscapes and "give expression to the enduring relationships of life" with an overlap of the mythic and natural at certain visible points. The boundaries of such local places, fixed in mythology, are not amenable to political negotiations.[33]

Polynesian Peoples

Polynesian peoples are found in the lands within the triangle formed by Hawaii, Easter Island, and New Zealand. To Polynesians all things, inanimate as well as animate, are endowed to some extent with *mana,* a power subject to nullification by human activity. Although Polynesian cultures have been eroded by Western influences, Western Samoa and Tanaa have been particularly successful in keeping intact their native cultures. Elsewhere an amalgamation has occurred.[34]

On Tahiti, Tangaroa made all the rocks and land, the roots and all that grows, seawater and freshwater, and fixed the dome of the sky on its pillars. At first the earth was atmosphere but later an affinity was established among the rocks of cliff and ocean, of slate, clay, and of pebbles and crumbling rock, all of which met and united. Then the land came rushing in. Hawaiki, birthplace of humans, became land through the chant of Tangaroa. Then roots spread upward and downward, inland and seaward, and the roots held the sand and the land became firm. "When Tangaroa saw all that was, he applauded."[35]

Hawaii's highly trivialized "aloha," or *aloha aina,* means love, reverence, and care for the land. To the native Hawaian religion is in nature with living gods and goddesses. Pele appears daily on the Big Island of Hawaii in the volcano and its lava, steam, and heat. Her family is found in the ferns, certain shrubs, and native trees. To the native Hawaian it is sacrilege to dig holes in Pele's body to capture her geothermal energy or to destroy her rain forests. In a very literal sense Pele is the land itself; that she is not through creating Hawaii is especially apparent on the Big Island.[36] The Polynesians are another people unable to conceive of religion and life being separable. Such separation was instilled by missionaries.[37]

The concept of Tapu—something not to be touched in any way—has been described as a religious observation for political purposes in that it strongly reinforces the authority of the priest or chief. It is a notable conservation measure, protecting such things as dwindling fish or coconuts. Reportedly, when Tapu was locally eradicated as mere superstition by Europeans, the results for such "resources" were calamitous;[38] yet another lesson of immediate consequences to be taken from a primal culture.

EASTERN TRADITIONS

Eastern traditions are as popular a source for reconnection to the natural world as are primal cultures. China is reputed to have an orientation "toward a harmony with the deepest rhythms of the universe." Through Taoism a "person attains an identity with every being in the universe." The Hindu world is reputed to be seeking a "transcendent realm of . . . Supreme Reality . . . beyond all knowing" with a feeling for the "unreality of the entire visible world."[39]

Language as a component of these cultures is relevant here, with terms of affection, for example, carrying "ultimate cosmological significance." The Chinese *jen* goes beyond love, benevolence, or affection and *ch'eng* goes beyond sincerity and integrity to be cosmic forces. Similarly, the Buddhist *karuna,* compassion, and India's *bhakti,* devotional love, convey the cosmological and spiritual.[40]

China and Oriental Traditions

The major religions of contemporary China are Buddhism, Taoism, Islam, and Christianity.[41] Throughout their imperial history the Chinese strove not only for cosmic harmony but for the proper enactment of rites, deemed synonymous with civilization itself.[42]

Buddhism

Buddhism is a pan-Asian religion and philosophy founded in northeastern India in the fifth century B.C.E. It spread into central and southeast Asia, including China and Japan. Siddhartha Gautama, the Buddha or Enlightened One, was a wandering ascetic in search of religious understanding and a way of release from the human condition. He instructed his followers in dharma, "true law," and the "middle way," a "path between a worldly life and extremes in self-denial."[43]

The Buddha's early teachings lead one to "the indescribable state of release called Nirvana" in part through accepting "the law of dependent origination." By the third century B.C.E., Buddhist doctrine and practice were formalized with an emphasis on the "composite nature of all things" and the "constant flux" of "phenomenal realities," "aggregates of momentary elements" lacking "any enduring selfhood." Thereafter new scriptures came into being with an expanded vision of the universe.[44]

Contemporary Buddhism is said to hold the universe and all its creatures in a state of complete wisdom, love, and compassion, acting in natural response and mutual interdependence from the beginning. Buddhism provides us with two particularly relevant lessons: the individual's place in the universe cannot be fully realized without giving up one's self, and ignorance is an obstruction to an effortless manifestation.[45] Through contemplation one can "pierce through to" the real nature of forms to perceive them as part of the great indivisible body of reality. "It cannot be cut into pieces with separate existences." Ultimately the individual's "life and the life of the universe are one."[46]

Confucianism

Confucious (K'ung-Fu-Tzu) was a contemporary of the Buddha. His teachings were more a social ethic than a religion in the ordinary sense. Confucianism provided substance of learning, a source of values, and a social code for the religion and philosophy of China, affecting Taoists, Buddhists, and Christians there and in Korea, Japan, and Vietnam. The foci are on *jen,* love, virtue, humanheartedness.[47]

Confucianism is chronicled in *The Analects of Confucious,* set down in the twelfth century. This is more a philosophy than a theology, because the search is for Truth, assumed to be in the singular and to be found in the *Analects* "hidden behind the words."[48]

Two passages are of interest here. One speaks of settling among the barbarians, to which a note adds that there is a frequent idealization of the noble savage in early Chinese literature and that when the Emperor "no longer functions, learning must be sought among the Four Barbarians," identified in the note as the cardinal directions.[49] The second attributes to the Master the observation that where nature prevails over the "ornamentation" known as culture, there is the "boorishness of the rustic"; when ornamentation prevails over nature, the result is "the pedantry of the scribe." The true gentleman arises from the two "duly blended."[50]

A contemporary observer notes that in Confucianism, the individual is not an isolated entity but part of a larger whole, "embedded in a supportive web" reaching from the family to friends. A personal commitment to self-cultivation contributes to the well-being of the larger society, through education as one example. According to the larger worldview, "rings of interconnection spread like circles rippling in a pond" outward "to embrace the political order" and beyond into the cosmos, a triad of heaven, earth, human.[51]

The philosophy of *ch'i,* material force, "reflects a synthesis of an enduring primal religious worldview of closeness to nature." Ch'i "asserts a common ground for all living things" and "provides a cosmological basis to understanding the identity and difference of . . . life-forms," while affirming the human role in the unfolding universe. Because each event, phenomenon, and individual is unique, each bears "moral value in the continually unfolding process of ch'i." Our task is to "understand and identify with the transformation of things" in a spiritual process. For some Confucian philosophers there is no distinction between ideal and physical nature.[52]

Taoism

Taoism is a religion and philosophy shaping Chinese life in association with Confucianism for more than two thousand years. Tao is described as an ineffable, eternal, creative reality; an absolute experienced only in mystical ecstasy. Once again there is emphasis on immanence and perfect harmony, here with one's original nature. But in embracing the simple and primitive and reducing selfishness, there is an antipathy to learning and sageliness.[53]

In contemporary terms Tao seeks the nourishment, preservation, and restoration of human nature, taking lessons from a simpler and more fulfilling life when nature and humankind worked together; *wu wei,* "do nothing contrary to nature." Tao is not so much the right way of life as the way in which the universe works, "the Order of Nature."[54] Each component has an inherent way of unfolding. The more effort exerted for control, the more disorder results,[55] an extremely valuable lesson for those who would interact with environmental amenities to any purpose, including protection.

Chinese Culture

The Chinese ideograph is derived from a drawing of the corresponding physical component of the environment. Each bears some strokes relating to humanity, nature, or deity. Significantly, the ideograph signifying prosperity includes all three, suggesting the impossibility of human arrogance in acknowledging that everything is connected with everything else.[56]

On a wholly different level, in China today there are practitioners of *feng shui,* literally "wind and water," the art of choosing locations for specific purposes through geomancy or the discovery of "harmonious relationships to their surroundings."[57]

Japanese Culture

The religions of Japan include Shinto, Buddhist sects, and Christianity. Contemporary Japanese culture is an inseparable mix of the traditional with introduced Chinese and Western cultures.[58] Shinto appears in Japanese literature late in the sixth century, following the introduction of Buddhism. The distinct Shinto way of *kami* is one of sacred power, polytheistic in nature. Although subject to the external influences of Confucians, Buddhists, and Tao, Shinto is closely associated with festivals and ceremonies and inseparable from governmental affairs. *Maturi goto* means the affairs of religious festivals, ultimately leading to government.[59]

In the Japanese story a male and a female deity created an island in the ocean from which they generated both the islands of Japan and the deities of nature, from mountains to crops. Japanese attitudes toward nature are reverent, friendly, and intimate. Mountain peaks and smoking volcanoes are viewed as collaborating spirits. Farmers perceive mountains as protectors and as sources of water needed for rice. The helpful kami are purified ancestral spirits dwelling in the mountains. "Kami" is not readily translated. They are omnipresent in natural beings and entities, leading to an "overarching reverence" distributed widely among such entities as the flower bud, "the veins and wrinkles on the tiny stone," Mount Fuji, even ideas such as growth or creation itself. There is no homogenization of kami; each natural entity has its own.[60]

In Japan the individual "is an inseparable aspect of the landscape," and it of the individual. Architecture engages in "conquest by surrender," building of wood with a strong emphasis on the horizontal, merging indoors and outdoors and minimizing boundaries. "The approaching visitor can see through the building to the garden on the other side." From within, open screens allow one to take in "the house-scape and landscape in a single sweep of the eye." The house is a part of the landscape, inseparable from the garden, a microcosm of design for all seasons, making the most of seasonal changes. As in Shinto, the reach is outward rather than upward.[61]

In times of turmoil or when life at court, or its equivalent, becomes constraining, the Japanese withdraw to a secluded place "in the midst of nature." They, too, are torn "between the splendor of the city and the Taoist-Buddhist allure of the countryside." They too are faced with city, suburb, and rural dichotomies and seek the Simple Life in its pastoral setting.[62]

Hinduism

Founded in the Vedic era spanning 1500 to 500 B.C.E., India has been influenced by Buddhism and Islam. The first Hindu empire arose in northern India.[63] Hinduism covers the whole of life, with Vishnu and Siva emerging as dominant at the juncture of the Common Era. The fundamental belief is "in an eternal, infinite, all-embracing . . . principle of ultimate reality called Brahman." Brahman is pervasive in all beings.[64]

Prince Rama of Ayodha is a hero of the Hindu tradition. Rama, dispossessed by his stepmother in favor of her own son, goes into exile with his wife and faithful half brother. When his wife is abducted by demons, Rama and his brother defeat them and rescue the woman with the aid of a monkey army. Returning to Ayodha, where his half brother was too righteous to take the throne, Rama accedes. Another hero, Sivaji, rebels against the ruler and becomes legend through his "ability to vanish into the forests and rocks." He later is crowned and establishes a lasting kingdom.[65]

Once again, to be whole and puissant, the Hero blends a civilized and wild self. India herself was experiencing the growth of urban life, emergent monarchies with extensions of power "engulfing the older, simpler life of the villages in vast and much less personal systems" rendering many an individual ill at ease and raising "basic questions about the nature and purpose of human existence."[66]

NOTES

1. Snyder, *The Practice of the Wild* (1990), 143.
2. Swimme and Berry, *The Universe Story* (1992), 153.
3. Roszak, *The Voice of the Earth* (1992), 93.
4. Tuan, *Passing Strange and Wonderful* (1993), 128.
5. Lopez, *Arctic Dreams* (1986), 273–274.
6. Pelikan, *On Searching the Scriptures* (1992), 75–76.
7. Pope John Paul II, *Crossing the Threshold of Hope* (1995), 78, 80, 81, 83.
8. Ninth-century Muslim philosopher Al Kindi (Faylasuf), from Armstrong, *A History of God* (1993), 173.
9. Ibn al-Arabi (Muid ad-Dinibn al-Arabi), twelfth- and thirteenth-century Muslim mystic, quoted in Armstrong, ibid., 239.
10. Quoted in Armstrong, ibid., 240–241.
11. Armstrong, ibid., 241.
12. Tuan, *Passing Strange and Wonderful*, 121, 122, 128.
13. Ibid., 161.
14. Jung, *Modern Man in Search of a Soul* (1933), 66.
15. Naylor et al., *The Search for Meaning* (1994), 128.
16. See, e.g., Clifford, *The Predicament of Culture* (1988), 213, 277 ff.
17. Berry, *The Dream of the Earth* (1988), 39.
18. Jung, *Modern Man*, 141, 142, 144.
19. Marsh, *Man and Nature* (1864, 1965), 39, note 36, 52.

20. Swimme and Berry, *The Universe Story*, 44.

21. Estés, *Women Who Run with the Wolves* (1992), 289.

22. Glavovic, "Traditional Rights to the Land and Wilderness in South Africa" (1991), 308, 309, 318, 319–320.

23. *Brittanica Micropedia* 1, Africa, 132.

24. Williams, *An Unspoken Hunger* (1994), 4.

25. Blier, "The Place Where Votun Was Born," *Natural History* (1995), 40, 42, 44, 47.

26. Wright, "Ecological Disaster in Madagascar and the Prospects for Recovery," in Chapple (1994), 11, 18–21.

27. Dorm-Adzubu et al., *Religious Beliefs and Environmental Protection* (1991), 3.

28. Ibid., 4, 16–17.

29. Ibid., 17, 19–21, 23.

30. *Brittanica Micropedia* 1, Australia, 712, 714.

31. Ibid., 714–715.

32. Devall and Sessions, *Deep Ecology* (1987), 12.

33. Lopez, *Arctic Dreams*, 296.

34. *Brittanica Micropedia* 9, Polynesia, 581, 582.

35. Alpers, *Legends of the South Seas* (1970), 57–59, 62.

36. Mander, *In the Absence of the Sacred* (1991), 319, 335.

37. Alpers, *Legends of the South Seas*, 33.

38. Ibid., 16.

39. Swimme and Berry, *The Universe Story*, 203–204. See also Tuan, *Passing Strange and Wonderful*, 187.

40. Berry, *The Dream of the Earth*, 20.

41. *Brittanica Micropedia* 3, China, 221, 229.

42. Tuan, *Passing Strange and Wonderful*, 187.

43. *Brittanica Micropedia* 2, Buddhism, 602.

44. Ibid., 603.

45. Devall and Sessions, *Deep Ecology*, 251.

46. Hanh, *The Miracle of Mindfulness* (1987), 47, 48.

47. *Brittanica Micropedia* 3, Confucious, 530.

48. Pelikan, *On Searching the Scriptures*, xv, 73, 74.

49. Ibid., Book V:6, 108, note 1.

50. Ibid., Book VI:16, 119, note 3.

51. Tucker, "An Ecological Cosmology," in Chapple (1994), 107.

52. Ibid., 113, 115, 121.

53. *Brittanica Micropedia* 11, Tao, 551.

54. LaChappelle, *Sacred Land Sacred Sex* (1988), 90–91, 95. See also Boorstin, *The Creators* (1992), 420.

55. Devall and Sessions, *Deep Ecology*, 11.

56. LaChappelle, *Sacred Land Sacred Sex*, 24.

57. Sheldrake, *The Rebirth of Nature* (1991), 178.

58. *Brittanica Micropedia* 6, Japan, 496, 497, 499.
59. *Brittanica Micropedia* 10, Shinto, 745.
60. Boorstin, *The Creators*, 136–137.
61. Ibid., 138, 144–145, 146.
62. Tuan, *Passing Strange and Wonderful*, 129–130.
63. *Brittanica Micropedia* 6, India, 285, 286.
64. *Brittanica Micropedia* 5, Hinduism, 935.
65. Cavendish, *Legends of the World* (1982, 1989), 16, 26–27.
66. Ibid., 32.

9

Images from Our First Nations

According to the Cherokee story, at the time of creation a white man was given a stone, an Indian silver. Both threw their gifts away as useless. Later the white man took up the silver as a source of material power, while the Indian revered the stone as a source of sacred power. Thus are represented the profound differences in the respective value systems of the European- and Native-American.[1]

Today in our search for personal and cultural articulation of our relationship with the natural world, we can be said to be seeking ways to perceive silver in a different light and to grasp reverence for the stone. As Native American poet Peter Blue Cloud asserts, the Amerindian individual lives in a continuity of past, present, and future, where the future is as important as the past because all three are connected. Where European-Americans see a straight line without curves or switchbacks, Indians perceive time in a circle.[2]

WHO ARE OUR FIRST NATIONS?

America's First Nations are many things to many people. For the record, they are first and foremost *themselves,* culturally and individually, of history and in the present. They were and are nations, cultures, *and individuals.* Nevertheless, Amerindians are also images — of the noble savage, wild man, or Wild Man — and representative of a way of life believed to be more intimately connected to the natural world than are we.

Native cultures are of interest here from two perspectives: that of potential influence on the emerging nation of their political and social structures and that of lessons from their relationship with natural phenomena. Here those perspectives are considered in historical context.

Significantly, we see much of Indian expression as art, but within native cultures there was no word for art. "The beautiful objects that people created were part of the integral fabric of their lives. While fine work was . . . appreciated, art for art's sake was unknown."[3]

Despite praise of Amerindian attributes from explorers as early as Columbus,[4] our historical interactions hold little promise for gaining insights. We have consistently found communication with other traditions difficult. Instead of experiencing "a shared human context" we have elected to confer "salvation" — political, social, economic, religious — on other cultures, whether they wanted it or not. At the same time we have "resisted incorporating the resources offered by others," with disastrous results resonating throughout the traditions of both "them" and "us."[5] Today we are living with the results, as are surviving members of America's First Nations.

If we have in fact reached a juncture in our interactions whereby we are actually "[r]ecognizing individual and group differences, and according these differences value and protection under the law," if we are prepared at last to denounce policies forcing "assimilation and subjugation" on First Nation cultures and individuals, then it may actually be that we are prepared to accommodate their interests while serving our own through what they can tell us.[6]

The prospect of our acceptance of the cultural values Amerindians have to offer is hardly novel. On no less than three profoundly important levels there are references to First Nation contributions to what makes us what we are. Jung claimed that European-Americans actually bear "an American Indian component in the[ir] psyche. Indian ideals of freedom have had their influence on our own federal Constitution. And modern dance owes much to Indian renditions of earth rhythms, with Isadora Duncan expressing such rhythms and movements in her early career.[7]

One point is worthy of note in this context: "[N]ative cultures that have lived successfully in one place for millennia have been abiding by successful economic practices, including wildlife and resource conservation." Significantly, both contemporary natural scientists and Amerindian cultures have engaged in an empirical form of ecology, but the latter have been at it for centuries in comparison to the ecologists' mere decades.

And, arguably, at least some First Nations intentionally underused resources in contrast to perceiving environmental amenities as economic resources, neglecting "direct knowledge of the environment."[8]

WORLDVIEWS

The foundation of discussion here is a matter of worldview — the vision of the world embraced by a culture and its members and its implications for their relationships to the environment, to others, and to the cosmos. Abstract and distant though it may be to the individual, that worldview forms the pattern of day-to-day living.[9]

The problem with most anthropological efforts to discover worldviews lies in the limitations of the anthropologist, who rarely "observes" a culture (mostly from an outsider's perspective) for more than a few years. Resultant writings are reduced to a few pages on "nearly infinite complexities" of a given culture. With a significance that extends to more than one level, the complex of a primal hunter-gatherer culture with nature can be compared to that of living cells interacting with their environments.[10] Both units — one of culture, the other of life — are beyond the capacity of contemporary scientists to describe fully.

For example, from Eskimo knowledge alone whole volumes on the ecology and behavior of arctic animals could be written. For thousands of years that "anecdotal" lore has been a matter of life and death. That kind of intimate knowledge can be attributed to native peoples throughout the North American continent.[11] It is instructive that the array of cultures described often share views of such universals as astronomy albeit distinct from the view of Western science.[12]

As for the popular view of Amerindians, it has been trivialized to the "murdering redskin" or the "noble savage." So perfunctory a reference neglects the very real potential for insight provided by the latter in a time when many suspect "the secret of surviving" may come from the First Nation past.[13]

Origin Stories

Worldviews are closely associated with a culture's origin stories that also inspire its traditions and arts. Origin stories reflect a people's aspirations and intended legacy. It is said that for Amerindians a "life's journey is an effort to keep the wisdom from the time of Creation a part of . . . everyday existence." Many believe creation is an ongoing process in concert with humankind as cocreators.[14]

In the Iroquois origin story a pregnant woman fell to earth where she was supported on the back of a turtle assisted by other animals. By magical means the woman grew into a mighty land. Her two grandsons became sources of earthly goods and earthly evils, the good creative, the evil disrupting or altering creations until banished by the good grandson.[15]

In the Lakota beginning the sacred and temporal dimensions were one. Consequently, a Lakota "still recognizes himself as a microcosmic reflection of the macrocosm." Life "in concert with the holy rhythm of that which causes all life to move" assists "the ongoing process of creation."[16]

Relationship with the Land

Native Americans are almost universally held to reject ownership of land in the sense intended by Europeans. Instead use of the land is the primary principle and that in terms of attention to the good of the larger community rather than of the individual. Land abandoned by one person was freely used by another, and not even a chief enjoyed any right to alienate regions traditionally used by a community.[17] Sacred Mother Earth is hardly alienable,[18] although there is some debate among the observers regarding ownership as such. It appears that New England's Algonquian sachems allocated territories to families or villages. Although European invasion altered Native practices, there is some indication that tenure in land shifted in response to ecological conditions, and conservation practices were followed.[19] In at least one instance, European influence led to a kind of conservation. In response to pressure for valuable beaver, Labrador's early seventeenth-century Montagnais people avoided depletion of the resource by establishing private hunting grounds and engaging in a kind of sustainable-yield management.[20]

Significantly, the Pequot War of 1637 and King Philip's War of 1675 have been attributed to the Indian attempt to end the greed for land manifested by the English.[21] In the southwest, the Pueblo and Navajo peoples live *in* the land and attribute no validity to any separation of people and landscape. Indeed, our contemporary estrangement is deemed unfortunate as well as false, one of the "great afflictions of our time," rejected by these people in favor of a strong "principle of harmony in the universe."[22]

The "Food Web"

Native foods are illustrative of a sense of balance or reciprocity with nature. As the key to survival, health, and even political and economic domination, food represented the "web of life interweaving native people" through festivals and ceremonies, rites of passage, healing rituals, and even so mundane an affair as changing diets. One element of the consequent integration with nature is identified as animism, a belief that "everything has a spirit and must be respected." While striking differences may occur among tribes, commonalities influenced by the lands of a people's range are observed.[23]

For Amerindians animals, among other beings, are not relegated to resource status reduced to quantification. Our ecologist's food web is for them a web of living systems expressed through teachings, stories, and ritual and integrated into social structure, status, and psychology.[24] For the Plains peoples, dreams and visions intensify relationships with animals. For the Oglala Sioux, cognition shifts beyond the phenomenal animal to archetypal essences in animal form.[25]

OBSERVATIONS OF POLITICAL STRUCTURE

If we are to return for a fresh look at what our predecessors on this continent can tell us, the initial interactions between our people and theirs are worthy of some review. It should at least begin with contemporaneous observers. Among them are Robert Beverly, who wrote *The History and Present State of Virginia* in 1705. He devoted a book in that history to the habitations and government of Amerindians, observing that they lived in "townships" of fifty to five hundred families, which groups were prepared to act collectively in emergencies. Beverly also observed the equivalents of a "viceregent," acting as governor, judge, and chancellor to the "king" who ruled over several towns. Thus, the precedent of government is established and compared favorably to European counterparts. From this model Beverly supported his position that the land held an "immanent garden" if his fellow immigrants "would adopt an ordered communal organization validated by nature itself."[26]

Beverly's was not the only description of Indian government and society of the prerevolutionary period. There was Lewis Henry Morgan's description of the League of the Iroquois and also Cadwallader Colden's 1727 *History of the Five Indian Nations*.[27] The vitality of native governments is suggested in the successful resistance of the Iroquois Confederacy to dispossession of land.[28] Later, Tecumseh sought a coalition of tribes against incursions of the United States by appealing to an Indian brotherhood among "children of the Great Spirit" who walked "in the same path."[29]

There was also mutual respect in these early encounters. Morgan, for one, made reference to the Iroquois's "position of authority" and "enduring institutions" and spoke of that people's "friendly relations" with both the Dutch and the English. The Iroquois, at least, demonstrated "singular fidelity" to a "covenant chain" terminated by independence of the States.[30]

REFLECTED IMAGES OF THE NATURAL WORLD

European reflections of how First Nations were interacting with the natural world commenced as early as encounters with Mayan civilization. It has been asserted that it is Mayan agriculture that led to the mysterious abandonment of their cities. Mayans engaged in cut and burn agriculture with an easy yield of maize from a few acres over a few years before the yield dropped and the farmer moved on. With time, the abandoned plots grew with tough grasses before reverting to jungle. Thus the cities were surrounded by plains the Mayan farmer could not cultivate in the absence of draft animals or tools beyond a pointed stick. When the cultivated plots became so distant city life was impractical, the people migrated and the jungle invaded.[31] Truth or pure imagination, there are lessons in such an explanation.

Thoreau observed of contemporaneous Amerindians, "Nature must have made a thousand revelations to them which are still secrets to us." Unfortunately, even that early Thoreau also had to take to heart the lament of a "wise old Indian" for the loss of "a great deal" of the lore of medicinal plants among the younger generation of his people.[32]

Perhaps the most valuable lesson provided by our First Nations is that of the complete integration of the material and spiritual realms. All components of culture are interconnected and the sacred is not separate but a permeating reality. Relationships are dominant and ubiquitous, in a series "always reaching further and further out." The resultant revelation is "I *am* the universe."[33] For such peoples the spiritual realm is inseparable from the economic and the political.[34] Their traditions wove "earthly spirituality into every aspect of life," the environmental "cause" is a "song of love," and the individual mission a restoration of "inherent harmony."[35]

Among some native peoples there was a "Golden Age" when people and animals talked with each other. These peoples came "to revere the sacred forces of the living, natural world." Like many of us, they long for "a dimly remembered time of wholeness" once "found in their highest rituals" and in "their everyday lives."[36]

For the Oglala "fluidity and transparency" in their perceptions of the phenomenal world preclude the drawing of any "absolute line" among worlds of "animals, men, or spirits." Yet that world is neither unstructured nor chaotic because of a binding principle: *Wakan* — "an ultimate coalescence of the multiple into the unifying principle of *Wakan Tanka*."[37] The phrase defined as the Great Spirit,[38] *Wakan* seems to go beyond "spirit" and to be more personally intensive than "holy," while *tanka* seems to mean a kind of unification of the disparate[39] — for Christians reminiscent of God as one spirit manifested in three "persons" and both transcendent and immanent.

Among California's coastal Chumash people were astronomers whose duties went beyond astrology and weather control into politics. To them, and many other peoples, our philosophical and scientific separation from nature would have been wholly foreign. Their basic assumption was that humans are "inextricably bound" with everything else in the world. Proper behavior was crucial to the continuing cycle of re-creation, of life coming from life. Prophetically, the Chumash are representative of the belief that animate matter is never destroyed but reconstituted in other forms.[40]

When such a people perceives a plant as a living creature, with skin, organs, and a soul, or as a female being, the source of life, that imagery is not merely metaphorical[41]: There is another world behind that we see. That other is the same one but "more open, more transparent, without blotches." There, as if within "a big mind, the animals and humans all can talk, and those who pass through . . . get power to heal and help. . . . To touch this world no matter how briefly is a help in life. People seek it, but the seeking isn't easy."[42]

This fundamental integration is reflected in small ways by numerous peoples responding in spatial patterns to cosmic rhythms: Southeastern peoples feed a sacred fire with four logs along the cardinal directions. The doorway of a Navajo hogan is oriented to the rising sun. The Mescalero Apache enters a tipi in a sunwise direction, and within, the Sioux moves in a sunwise direction, east through south to west. A Hopi priest prays to the four directions, moving from west to east. The Pawnee sing of linkage to a guiding constellation (Pleiades). Roofs of Pawnee earthen lodges are supported by four posts representing the "stars of the four cardinals."[43]

Upon completion of a hogan, a Navajo prayer exhorts life in harmony with nature, "participating in the cosmic plan laid down by the gods in the beginning of time. Living in beauty is also to live in accordance with the rhythms of the heavens." They and the Pueblo hardly distinguish sacred from secular, deeming life sacred and living it "right" — itself a sacred act. "Those who discover the means of establishing and maintaining order hold the key to life's processes."[44]

The Navajo are known for their healing through sand paintings. The practice arises from the belief that illness results from a world at odds with itself and is a part of the cosmos. Thus, the illness is overcome when the sufferer accepts that responsible "tensions and oppositions can be balanced in a unity" signifying "good health and beauty." The sand painting process is a "cosmic map" for restoring health and beauty in both life and the world, with the suffering flowing away through the painting. As the healing sufferer acknowledges "the fullness and unity of the reality" represented by it, the picture disappears in concert with "the dissolution of the tensions and imbalances which have given rise to the suffering."[45]

TRANSITION FROM THEIR PAST TO OUR PRESENT

By the nineteenth century we were in the final process of our "conquest" of this continent. Coincident with this process was the "formal creation of American literature." With it images of Amerindians had been in evolution from heathen to savage to the juxtaposition represented by the near-archetype, noble savage, all of them deemed at best an insult by at least some of today's Indian writers. As Ward Churchill proclaims, the mythic cannot be "murdered, expropriated and colonized" in the so-called "real world."[46]

When we pressed Amerindians to follow our lead, the holistic perception of the cosmos was suppressed to the "grave injury" of traditional native vitality in favor of what is emerging today as our "less rational approach."[47] The Lakota tell us they continue to "sense that the ways of the Great Spirit are a mystery" compelling humankind "to continue their search in both the physical and the spiritual worlds." The reward is new insight and "an awed response to mystery. If one gives one's entire being over to exploring the patterns and possibilities in the environment, then new possibilities are always in the making and one is caught up in a forward momentum."[48]

The Lakota "vision quest was a journey to the inner, spiritual landscape that showed the quester the direction to follow in the travels through the physical landscape."[49] A Hunkpapa Lakota medicine man who still speaks Lakota and whose father and older brother are Christian clergy assures us "I can talk both ways, Indian way and Bible way; if you *really* know about them, they are the same."[50]

NOTES

1.   Nabokov, *Native American Testimony: A Chronicle of Indian-White Relations from the Prophecy to the Present, 1492–1997* (1978, 1991), 32.

2.   Bruchac, *Survival This Way* (1987), 40.

3.   Lester, "Art for Sale," in Weinstein (1994), 151.

4.   Fairchild, *The Noble Savage* (1928), 9.

5.   Berry, *The Dream of the Earth* (1988), 182.

6.   See, e.g., Duthu, "Implicit Divestiture of Tribal Powers: Locating Legitimate Sources of Authority in Indian Country" (1994), 357–364.

7.   Berry, *The Dream of the Earth*, 188.

8.   Mander, *In the Absence of the Sacred* (1991), 257, 259, 260.

9.   Heizer and Elsasser, *The Natural World of the California Indians* (1980), 204.

10.   Kellert and Wilson, *The Biophilia Hypothesis* (1993), 205.

11.   Ibid., 207–208.

12.   Williamson, *Living the Sky* (1984), 299.

13.   Bruchac, *Survival This Way*, x–xi.

14.   Hill and Hill, eds, "Creation's Journey" (1994), 30. Excerpts from inaugural exhibit and book opening New York's National Museum of the American Indian.

15.   Ibid., 31.

16.   Amiotte, "The Road to the Center," in Dooling and Jordan-Smith (1989), 246.

17.   Williamson, *Living the Sky*, 29.

18.   Kavasch, "Native Foods of New England," in Wienstein (1994), 12.

19.   LaFantasie, *The Correspondence of Roger Williams* (1988), 244. See also Anderson and Leal, *Free Market Environmentalism* (1991), 128.

20.   Anderson and Leal, ibid., 65.

21.   Weinstein, *Enduring Traditions* (1994), 53–54.

22.   Momaday, in Bruchac, *Survival This Way* (1987), 179–180.

23.   Kavasch, *Native Foods of New England*, 5, 11–12.

24.   Mander, *In the Absence of the Sacred*, 260.

25.   Brown, "The Bison and the Moth," in Dooley and Jordan-Smith (1989), 178.

26.   Machor, *Urban Ideals and the Symbolic Landscape of America* (1987), 78–79.

27.   Ibid., 79.

28.   Gaddis, *American Indian Myths & Mysteries* (1977), xiv.

29.   Nabokov, *Native American Testimony*, 96.

30.   Morgan, *League of the Iroquois* (1993), 10.

31.   deCamp, *Lost Continents* (1970), 108.

32.   Thoreau, *A Week on the Concord and Merrimack River — Walden, or, Life in the Woods — The Maine Woods — Cape Cod* (1985), 731, 774.

33.   Dooling and Jordan-Smith, *I Become Part of It* (1988), 20.

34.   Mander, *In the Absence of the Sacred*, 208.

35.   McGaa, *Mother Earth Spirituality* (1990), xiii.

36.   Nabokov, *Native American Testimony*, 50.

37.   Brown, "The Bison and the Moth," 181–182.

38.   Jabner, "The Spiritual Landscape," in Dooling and Jordan-Smith (1989), 193.

39.   Brown, "The Bison and the Moth," 177–182.

40.   Williamson, *Living the Sky*, 277, 279–280.

41.   Mander, *In the Absence of the Sacred*, 211.

42.   Snyder, *The Practice of the Wild* (1990), 164.

43.   Williamson, *Living the Sky*, 298.

44.   Ibid., 153, 155–156.

45.   Gill, "Disenchantment," in Dooling and Jordan-Smith (1989), 77.

46.   Churchill, *Fantasies of the Master Race* (1992), 34, 38.

47.   Mander, *In the Absence of the Sacred*, 260.

48.   Dooling and Jordan-Smith, *I Become Part of It*, 201.

49.   Ibid., 202.

50.   Mattheissen, *In the Spirit of Crazy Horse* (1991), xxix.

10

Contemporary Amerindian Perspectives

The tree of life occurs in numerous cultures as a symbol of the sacred nature of creation and of the bonds connecting the material and spiritual realms. The tree also reflects the continuation of life and peace. Through their beauty as well as in their roles as sources of oxygen, food, shelter, and building materials, trees are a natural expression of interconnectedness. The Great Tree of Peace also is the symbol of the Iroquois Confederacy.[1]

That American First Nation images continue to affect us is informative. Novelist Tony Hillerman recalls his initial image of native peoples, formed out of his encounter with a group of Navajo participating in a healing ceremony. Hillerman was driving on a dirt road when "a group of mounted men and women emerge[d] from a thicket" to cross ahead of him. He could see no trail; "the riders seemed to emerge as if by magic" from a sandstone cliff. Some wore ceremonial attire.[2] Hillerman's novels of contemporary Navajo detectives are suggestive of just how evocative that encounter was.

The potential for rediscovering what our First Nations can tell us of being connected to the natural world in mutual respect and benefit is appealing. But what of contemporary Native American cultures and individuals? What are their circumstances?

THE LIFE OF CHARLES A. EASTMAN

Charles A. Eastman is in some ways the emotional ancestor of contemporary Amerindians, whether of one or mixed blood. Eastman's maternal grandfather was a Caucasian artist, but Eastman himself considered himself Santee Sioux, Ohiyasa, the last of five children. When the starving people of his Montana reservation rebelled in 1862, Ohiyasa was taken to Canada by his grandmother and uncle, where he was raised to be a warrior and hunter. He grew up hating European Americans for the apparent death of his father.[3]

Before Ohiyasa's initiation ritual, his father, now a Christian, reappeared and the boy was returned to Dakota Territory. He was to recall his traditional life with affection; it was already gone when in 1902 he published a book recounting it. Over the years Ohiyasa served as a mediator between the conflicting cultures, became a Christian, and earned a medical degree. In his early thirties he returned to South Dakota's Pine Ridge Agency as a government physician. Over the years he held that and several other positions with the Bureau of Indian Affairs.[4]

His initial response to school did not compare well to his previous nature studies. The adult Eastman calls his younger self "a wild cub caught overnight, and appearing in the corral the next morning with the lambs." Ultimately, Eastman, having gone into the forest to seek the "Great Mystery" in silence, became determined "to follow the new trail" without relinquishing his people's traits. Indeed, his father believed that "Great Mystery" had shown both peoples "the good and evil, from which to choose" and taught the same virtues to both peoples.[5]

Accepted by both cultures, Eastman tried to bring them to mutual acceptance. But it is asserted that he must have faced tremendous "inner pressures and conflicts." He urged Amerindians to accept the positive aspects of European American society, but, though he was able to reconcile his two heritages, his "views on the compatibility of Christianity and Indian religious training are controversial." He retained much of his spiritual identification with his own race and sought to help Indians adapt to European American views while retaining their Indian identity.[6]

Eastman must have had some effect on his European American audiences. He records that some among them "admitted that morality and spirituality are found to thrive better under the simplest conditions than in a highly organized society, and that the virtues are more readily cultivated" when one struggles against "the forces of nature" than when the struggle is "with one's fellow-men."[7]

CULTURAL AND INDIVIDUAL PERSPECTIVES

Theologian Niebuhr declares that sin is separation from the truth. If our First Nations' holy visions are truth and that truth is interconnectedness with all of existence, then alienation is sin. "People living in the modern age have become alienated, feeling alone, isolated and separated."[8] We look back upon five centuries of invasion to perceive the First Nations in retrospect as among the freest of peoples. Today they are confined, culturally as well as physically.[9]

Yet we turn to them among an array of ancient ones "trained in the sacred traditions of this land" as we "experience the first tentative pangs of a new culture."[10] But these peoples, too, are experiencing ambivalence and a degree of alienation from the natural world. While "hemmed in by concepts of continuous tradition and unified self" their identity with their own cultures "must always be mixed." The scholars among them who are "[i]ntervening in an interconnected world" are "caught between cultures." In fact, even any who wish to immerse in traditional ways may find "no distant places left on the planet where the presence of 'modern' products, media, and power cannot be felt."[11]

Exile to an Amerindian bears a religious dimension. Vine Deloria's recollection of the archetypal exile is well worth close attention: Such exile arises from "expulsion of the chosen one from his comfortable and often exalted position in society." The exile is "thrust into a barren place where he has to abandon his former knowledge of the world." Here the exile "learns humility and faith, comprehends the transcendent nature of ultimate reality, and is initiated into the mysteries and secrets of the other, higher world." Armed with this enhanced knowledge, the exile returns to his society to create "fundamental and lasting reforms, so that society marks its distinctive identity from the time he received his exilic commission."[12]

One parable tells of a man of magic who did not know the identity of his father. This man fooled all the people who became so caught up in his magic, "they neglected the mother corn altar." Failing to recognize it to be a trick, they came to rely on the new magic. Mother Nau'ts'ity'i grew angry and left them to live off the new magic and took with her "the plants and the grass." No new animals were born; she even took the rainclouds away with her.[13] The lesson here is plain.

The same warning is told a different way in the historical prediction of Luther Standing Bear in the early part of this century. This Sioux author warned that the European invader "cannot live in peace until he comes to understand and love this land" and sees the land itself "to have a measure of dignity and respect." Until that day those "who live on the land are . . . incomplete."[14]

On the Land

Land is indeed "the absolutely essential issue defining viable conceptions of Native America." It is a "deeply held sense of unity with particular geographical contexts" that once provided and continues to afford "the spiritual cement allowing cultural cohesion across the entire spectrum of" First Nation cultures.[15] To the Native American, land is a gift to a tribe from higher powers. Tribes in turn assume "certain ceremonial duties which must be performed as long as they live on and use the land." This is the dimension that has been stolen. This is the real exile. And our society has failed "to offer a sensible and cohesive alternative." Consequently, individuals became disoriented. "They could not practice their old ways, and the new ways . . . were in a constant state of change" because they were not cohesive.[16]

The interconnection between Native people and land as Mother is no mere romanticism. Land is "a cultural centerpiece with wide-ranging implications for any attempt to understand reservation life." The mixture of subsistence, spiritual origins, and sustaining myth establishes a "landscape of cultural and emotional meaning" that can be determinative of "the values of the human landscape." This is a cultural taproot that has been nearly severed over the last three centuries, but it is a root that "is being rediscovered and tended with renewed vigor and stewardship." It is a recurrent theme in contemporary Indian literature. Today reservations are the places "where the land lives and stalks people," where it "looks after people and makes them live right"; it is the place where "earth provides solace and nurture."[17]

Occasionally right prevails. In 1879 the Narragansett tribe was terminated by the State of Rhode Island. In 1883 an individual averred that the same blood was still "running through our veins," leading to pressure from the 1890s to the 1930s for the return of the lands. In the 1970s a lawsuit was settled with the return of a few acres. But the "Narragansett homeland endures." In the 1980s, Rhode Island "formally acknowledged return of lands." One Narragansett called one's "small portion of mother earth" the most valuable thing, "a sacred thing when you walk the land, and you drink from a spring, and you hunt in a forest, [as] for a thousand or two thousand years or more your direct ancestors have done."[18]

Not all Native peoples are so fortunate as the Narragansett. A member of the White Mountain Apache Tribe laments that the "children are losing the land. It doesn't work on them anymore. They don't know the story about what happened to those places." Contemporary Lakota note that their personal relationships are defined by their relationship to the earth. Their problem is "*how* to get back in the relationship to the earth." The relationship has lapsed from active to passive. Equally disturbing is the conjuring of "disturbing utopic visons that endlessly romanticize the people and the land."[19] And yet utopia and romantic notions may reflect a first tentative step in the right direction — if neither perverted nor converted into yet another kind of taking.

Poet Paula Gunn Allen echoes the sentiment in discovering through another's novel something that brought her land back to her. Her personal sense of loss was in fact "land sickness — loss of land. It brings back grief to you not to be at home."[20]

On the Residents and Denizens

Animals and other denizens of the land are important sources of connection as well. Zuni tradition promises the presence of the spirit of the beast represented in animal fetishes or charms, once blessed. The Zuni themselves draw upon these powers and even the occasional European American acknowledges their efficacy as basic earth medicine. Each beast stands for something: The bear is symbol of solitude and introspection, the buffalo abundance; the mountain lion is assertive, the hawk a swooping predator, and the eagle represents a sweeping overview. To us these charms can represent the linear individual's hunger for the intuitive side and personal creativity. One dealer in the charms notes their role in reconnecting with the earth through an "inner journey." As a result the Zuni produce some fifty thousand of the charms a year and have voiced no objection to their use for "spiritual direction."[21]

In the Matter of Language

Angus Cockney Kaanerk, an Inuit artist, attributes his self-image to nearly two months of ski touring to the North Pole. His image is a mixture of Inuit and external traditions, a duality describing most of his work and placing him at the fringes of "the Arctic fraternity of carvers." As an observer notes of those who have forsaken their nomadic life for subsidized villages, "the new art made by people singing of what they no longer are contains an element of tragedy and irony." Kaanerk "sings of stone of his modern Inuit experience. Pitting the traditional

values of resilience, endurance and stoicism against the dislocations of a
fast-changing world, he captures modern themes."[22]

One of Kaanerk's pieces is particularly affecting. It is the tragic mask
of an unfinished grieving face. In English the caption reads "I can't speak
my language." The poignant image is drawn from the artist's own experience. He once had to ask another Inuit to speak English. Thus, his grieving man is lamenting more than his own people's losses. It is for us all,
who are cut off from our roots by technology-driven change.[23]

This loss of language, attributable to a combination of assimilation —
voluntary and otherwise — and a tradition mostly oral, is devastating. For
peoples who have to survive in two environments, there are losses in both,
with language remaining the heart of something precious. It is reassuring
to be advised that efforts are in progress to preserve Native languages.[24]

Arapaho Stephen Greymorning rewrote Felix Salten's *Bambi* in his
own tongue in rhyming verse. Disney produced it as a home video. Not
only does such a project make explicit the comparable cross-cultural
themes of speaking animals; it also serves the preservation of a language
on the verge of extinction.[25] Thus, it seems both cultures can benefit from
the recognition of lessons and linkages.

There is an impression that individual Amerindians, even living in
urban settings, retain some "indestructable reality" of "their own psychic
world." This other "mode of consciousness" requires a language going
beyond foreign into the mysterious. And its translation requires an insight
beyond any linquistic skill. Again, the suggestion is that this image is
beyond some merely romantic fantasy as a reality of abiding influence on
the continent.[26]

The Holistic View

Among the Micmac of northern New England and Canada's maritime
provinces, everything in the universe has a particular, conscious power
remaining in its possession until death, whereupon another animate being
emerges. Thus, the old power is not lost but transformed. A tree, for one,
becomes dead wood, but that wood is shaped and lives again with the
power of that shape and function. The animate die and remain dormant
until put to new use.[27] The connection to both natural and social cycles is
reassuringly obvious.

Clearly, to native peoples the world is not of disconnected, separate
pieces but a whole. While each person is an entity, none is separate from
everything else. This connection is not limited to our First Nations or to
this continent or even this planet. Moreover, all of us "are originally

tribal" although we "seem to feel separated from that" or to have forgotten.[28]

Visions among Indians and those of mixed heritage inform individuals of who they are, "how much Indian they are." White Deer of Autumn (Gabriel Horn) writes of "gazing up at the stars" and remembering a longing to touch them. He felt close enough to do so and to be a part of them. His vision also contained dolphins, of whom he felt a part, leading him to observe the mystery and wonder of being a part of everything, as we all are.[29]

As we seek our own enlightenment, we must remember we are dealing with peoples' present, not merely some romantic past. Tribal memory is a living and potent force affecting the future. Also worthy of retention is that "the center of tribal languages often has nothing to do with things, objects, but contains a more spiritual sense of the world."[30]

Today New England Native peoples have become nearly invisible to us. Yet one woman of the Schaghticoke Nation reminds us that her reservation in Connecticut has existed for two hundred fifty years. It is on ancestral homelands of more than eight thousand years' duration. She laments the historical silences contributing to her people's invisibility.[31] She reminds us that the time is upon us when the invisible are gaining increasing visibility. Those of us who would expand our notions of stewardship must hope for enlightenment in the company of visibility.

Place is a crucial element in enlightenment, including that of the spiritual realm. An Amerindian poet tells of seeking "that exact hollow where terror and comfort meet" as she returns to the "endless mesas" of her home. She speaks of the homely details of thistles, "bright bits of obsidian," and fragments left by old potters. She is seeking a "sense of being securely planted," in the language of the therapist of being "balanced but not grounded." The hollow is a metaphor of more than one level; it is the hollow of a hand, the hollow of the Lord's hand in Judeo-Christian tradition. This poet's work is replete with movement between "Christian awareness and a more Indian sense of land and place."[32]

Joy Harjo describes Anchorage as a city "of stone, of blood, and fish" and connects it to the mountains, whale, and seal. She takes the city back in time to the "glaciers who are ice ghosts" who carved the earth and shaped the city. Returning to the present she speaks of the "cooking earth" beneath the concrete and of the ocean of air above with its dancing, joking spirits, filling themselves "on roasted caribou" as "the praying goes on, extends out."[33]

DUALITY AND AMBIVALENCE

Notwithstanding the traditional holistic view of First Nations, the contemporary Amerindian suffers from the same sort of ambivalence we are experiencing. It is a mixed ambivalence of whole cultures and of individuals, which for them may have emerged early in this century when European Americans tended to place Indians in one of two categories: those resistant to change and those assimilating into the larger society. Clearly, the reality was hardly so clear-cut. One man, Carlos Montezuma, a Yavapai of southern Arizona, is representative of what was to come. To some extent he assimilated voluntarily and became a medical doctor who advocated assimilation. His mixed feelings, however, are reflected in his disdain for Indian cultures in combination with a deep belief in Indian abilities, his scorn for reservations he deemed prisons, and the fact that he despised the Bureau of Indian Affairs. Eventually Carlos Montezuma took up the resistance of his people and came to see reservations as homelands to be freed from the BIA. In the end he went "home" to die "in a brush shelter."[34]

Poet Linda Hogan began to write while in suburban Washington, D.C., because the split in her life's cultures "became a growing abyss" she would weave back together. Through her poetry she finds her way to "return and see things." She refers to "a life deep inside" that persists in asserting itself. "It is dark and damp, the wet imagery of my beginnings. Return. A sort of deep structure to myself, the framework. It insists on being written and refuses to give me peace unless I follow its urges."[35]

Another writes that being half-Yaqui is not easy. One side demands belief "that trees and rocks and birds talk" while the other demands belief "in glass-walled elevators and voices that are transmitted from space." She speaks of another who had tired of "shape-changing: being Indian with Indians and white with whites." The situation is reminiscent of Yomumuli, Enchanted Bee, a spirit woman of long ago, who left her people when she was not believed. When she went, she did so "with her favorite river rolled up under her arm, . . . her feet like two dark thunderclouds."[36]

Those individuals who must live in the city are "like enchanted trees, with bones for branches and eyes for leaves. If they listen, they can hear the humming of stars and bees and waves. . . . They are my ancestors, my future."[37]

Paula Gunn Allen tells of the mountain near her home, which is wilderness. But her place is between civilization and wilderness, posing the choice of direction. Her resolution is neither, but the middle, even as she values the wilderness aesthetic and its moral and personal value while accepting the personal usefulness of civilization. But she does judge civilization in terms of wilderness rather than the reverse. She notes in her poetry a "haunted sense" and "a sorrow and a grieving . . . that comes directly from being split, not in two but in twenty, and never being able to reconcile all the places that I am."[38]

A MERGING? LESSONS FOR RECONNECTION AND AN ENVIRONMENTAL ETHIC

These native peoples are not alone. They are speaking of themselves, but for us as well. We, too, must constantly stride in two very real yet very different worlds, if not realms. For all the hazards and all the alienation on so many levels within and between our peoples, at least one Amerindian sees "the American Indian traditions as fitting quite easily into the world-wide traditions."[39]

For better or worse, there is a history of such exchanges. Each of us has something to offer and something to gain, not just across the European and Amerindian line, but among Indians themselves. A willingness to adapt selectively led "to localized pan-Indianism in New England." With time, as "more coherent cultural wholes" emerged and New England Native peoples "achieved the recognition they so badly wanted, they abandoned [pan-Indian] practices in favor of their own traditions."[40]

There is a mixture of positives and negatives and there is ambivalence — among the artists, among those who appreciate the art, among the spokespersons for Native peoples. All of this is a metaphor for the larger cultural situation, with its longings and its endless potential for perversion. On a more positive note, there are those Native people who believe "[w]e are indigenous people to this land. We are like a conscience. We are small, but we are not a minority. We are the landholders, we are the landkeepers."[41]

Certainly, it is up to those who believe a sharing among cultures can enrich both participants in the process to refrain from any such usurpation. Sharing should always enrich, never impoverish.

Our conception of law is not the answer: We have many laws and consistently break them. Chief Frank Fools Crow advises us that his are a people of only one law and it is God's law that we "live on this earth with respect for all living things." Elder Matthew King adds the importance of respect for all living things. The heart, mind, soul are mingled with the world. He tells us "these hills are our church; the rivers and the wind, and the blossoms and the living things — that is our Bible."[42]

Conscience does not work when it is not allowed to speak out or it is not heard — and heeded. Are we, as Jung predicted, on the verge of spiritual rebirth with a fundamental change in our values and attitudes? Can our paradise be restored to us through a belief in science coupled with "a religious attitude"?[43] Perhaps it is time to embrace as our own elders "the undivided family of" humankind. Thus "those of us who live in the seemingly rootless modern metropolis might yet reclaim a universal ancestry that opens the myths, traditions, and teachings of the entire race to us."[44]

NOTES

1. Rockefeller and Elder, *Spirit and Nature* (1992), 12, 16.
2. Hillerman, *Hillerman Country* (1991), 34.
3. Eastman, *From the Deep Woods to Civilization* (1916, 1944), vi-vii.
4. Ibid., vii–ix.
5. Ibid., 23, 26, 28.
6. Ibid., xii, xiv.
7. Ibid., 188.
8. McGaa, *Mother Earth Spirituality* (1990), xv.
9. Berry, *The Dream of the Earth* (1988), 181.
10. McFadden, *Profiles in Wisdom* (1991), 12.
11. Clifford, *The Predicament of Culture* (1988), 10, 11, 13–14.
12. Deloria, "Out of Chaos," in Dooling and Jordan-Smith (1989), 260.
13. Silko, *Ceremony* (1977), 48–49.
14. Nabokov, *Native American Testimony* (1991), xviii–xix.
15. Churchill, *Fantasies of the Master Race* (1992), 131.
16. Deloria, "Out of Chaos," 261–262, 265.
17. Pommersheim, *Braid of Feathers* (1995), 14–15.
18. Weinstein, *Enduring Traditions* (1994), 85, 87.
19. Pommersheim, *Braid of Feathers*, 33–34.
20. In Bruchac, *Survival this Way* (1987), xiii.
21. Winokur, "Pushing Their Luck: Zuni Indians Peddle 'Magical' Charms," *The Wall Street Journal*, April 28, 1993, 1.
22. Ulman, "An Inuit Artist's Self Portrait in Stone," *The Wall Street Journal*, July 14, 1993, A10.
23. Ibid., A10.
24. Henry, "Endangered Speech: Native Americans Attempt to Save Tribal Languages for the Next Generation," *Christian Science Monitor*, June 29, 1993, 4.
25. "In the News," *Native Peoples* (Summer 1994), 94.
26. Berry, *The Dream of the Earth*, 191, 193.
27. Weinstein et al., "The Use of Feathers in Native New England," in Weinstein (1994), 182–183.

28.   Harjo, "Anchorage," in Bruchac (1987), 92.

29.   "Vision, Identity, and the Great Mystery," excerpted from "A Circle of Nations: Voices and Visions of American Indians," in *Native Peoples* (Fall 1993), 51.

30.   Harjo, "Anchorage," 94.

31.   Richmond, "A Native Perspective of History," in Weinstein (1994), 103–104.

32.   Allen, "Recuerdo," in Bruchac (1987), 3–5.

33.   Harjo, "Anchorage," 89.

34.   White, *"It's Your Misfortune and None of My Own"* (1991), 439–440.

35.   Hogan, in Bruchac (1987), 122–123.

36.   Endrezze, "The Humming of the Bees and Waves," in Leslie (1991), 78.

37.   Ibid., 81.

38.   Allen, "Recuerdo," in Bruchac (1987), 7, 18.

39.   Allen, *The Politics of Wilderness Preservation* (1982), 16–17.

40.   McMullen, "What's Wrong with this Picture?" in Weinstein (1994), 144.

41.   Lyons, "Our Mother Earth," in Dooling and Jordan-Smith (1989), 274.

42.   Mattheissen, *In the Spirit of Crazy Horse* (1991), 519.

43.   Jung, *Modern Man in Search of a Soul* (1933), vii–viii.

44.   Roszak, *The Voice of the Earth* (1992), 87–88.

11

A Sense of Place

We each have *a place,* to which we may be born, which we may find, or which we may "realize after long searching" to be the place left behind. Whatever our relations to that personal place, "it is made a place only by slow accrual, like a coral reef."[1]

WILDERNESS

Pollution may invoke the call to combat with a phalanx of recycling bins the coat of arms, but wilderness is the fundamental image underlying the evolving environmental movement. More to the point, wilderness is the environment unmarked by civilized hands.

Wilderness is the primary image motivating our personal call to stewardship. The image may be as trite as the tourist mistaking for a wolf the coyote scavenging in the garbage or as profound as the thinker immersed for a lifetime in the realities of wilderness. Wilderness is where our hearts — and relevant archetypes — reside.

As place wilderness harbors a variety of ideas, depending on personal interest. Wild areas hold something for science, ecology, recreation, philosophy, ethics, politics, and polemic. The ideas of wilderness can serve any combination of those interests, but wilderness is still founded in some part on what the individual "mind wants to make of it." To one it represents freedom, health, hope, and companionship.[2]

To another wilderness is a romanticized manifestation of the merger between the material and the spiritual realm, far from the outdated Cartesian model, a place where wild animals, alert and free, move "with a beauty beyond the predictability of a machine, far beyond the lock-step" of merely rational conclusions. Wilderness is beauty seething with intelligence, "surprising and refreshing for the human mind." That "mutations are a fundamental dynamic in the first cells of Earth" reveals "the wild freedom to wander" "rooted at the core of life," a freedom "to grope, to change spontaneously, to run galloping as an animal." In this wilderness mutations represent "the discovery of an untamed and untamable energy at the organic center of life."[3]

To yet another wilderness represents in retrospect the distinction between humankind and nature that attended the advent of farming and herding in our distant past. In prospect wilderness may lie on a continuum, one extreme "little more than a romantic anachronism," the other "bound up with the emergence of an evolutionary viewpoint on cosmological process." On that continuum the line separating civilization from wilderness may not be so firmly established as ordinarily perceived.[4]

We must consider our relationships with wilderness and nature at large: "Nature is born wild and should remain forever so." To dwell in society, we who arose from a wild nature "must succumb to civilizing." But for all our surface civilizing, a "residue of wildness in us responds to the wildness in nature's world" causing us to "seek and long for wildness." That is sufficient "reason to try to preserve the cathedral of the wild." Our world is sterile and empty without this wild spirit remaining at large. "Is it not worth our best efforts to save?"[5]

Wallace Stegner is one asserting that our wilderness helped form the national character as it shaped our history. Indeed, we will have lost something "as a people if we ever let the remaining wilderness be destroyed." We need to preserve wilderness as the challenge against which our character was formed. To be reminded and reassured of its continuing presence "is good for our spiritual health even if we never once . . . set foot in it."[6]

CITIES, SUBURBS, AND THE SIMPLE LIFE

While our pioneering ancestors were carving the garden from the wilderness, they were also seeking to synthesize cities from nature and blend the two. Ambiguity is immediately evident in those "denouncing Europe as an overcivilized, effete society lacking the resilience and vitality of its transatlantic offspring." Although Europe might have been richer in civilization's amenities, in America "the ideal of urban-rural synthesis was highly effective in" identifying a "healthy blend of city and country" for "the best of nature and the best of civilization." There was no need to renounce either. The pioneering architects of American cities thrived "under the banner of urban pastoralism," no oxymoron perceived. Pierre L'Enfant, architect of Washington, D.C., designed that city as a "compleat garden."[7]

Indeed, our ability to make our way in forests serves us in our cities. The forest teaches the lesson of reliance on our instincts and intuition. The city in its complexity demands such skills, because being alien not natural, it is "more of a forest . . . than the forest itself."[8]

A recent publication of the contemporaneous art of John Rubens Smith reveals him as a "visionary collector" who foresaw the rise of an industrial nation ending our agrarian ideal. An 1828 watercolor of the nation's Capitol is representative. The Capitol is surrounded by lawns and forest with cattle grazing in the foreground. This agrarian foundation is set off by a looming storm. There is every reason to believe that there is no accident in the composition. Smith's body of work is praised from the distance of time for its "encyclopedic breadth and minuteness of detail in a pivotal era."[9]

Emerson urged that cities "express the latently inherent relationship between cityscape and landscape." More than touches of green and of rural virtue are called for. The ideal affords equal value to the two components and maintains a harmony between them. After all, "the native terrain has provided a unique opportunity to create on a national scale an urban pattern that is one with the cultivated landscape." From England to America and from century to century, the overall impression to be conveyed is one of "peace, harmony, and continuity, with the city nestling in the natural setting as though growing from it."[10]

Frederick Law Olmsted has been identified as this country's "progenitor of landscape architecture." Olmsted's urban park planning "would profoundly affect subsequent efforts to make American cities more natural." His work is the art of one "who reveals the 'beauty and designing powers' of nature 'to cultivate the intellectual' in" us. His own fondness for nature arose from his childhood rural experience. Olmsted believed his "romantic love for cultivated nature and rural life" to be the foundation of his belief in the regeneration of "the human spirit in the city" through similar conditions. Thus, public gardens and urban parks would "alleviate" a city's "injurious conditions" and make an "urban world more humane and civilized" by inspiring "'harmonious cooperation'" among members of the community. In short, the goal is to create a nation of pastoral cities.[11]

In the first half of the nineteenth century, Baltimore could be described without apparent irony as a fairyland, a pastoral paradise. As for New York City, nature had done everything for it; no dark alleys of confinement or noisome atmosphere spoiled it.[12] What was lacking was a hero or any "urban counterpart to the noble husbandman to seize the imagination." The transcending question was (and is) whether a hero could be created from the American character, "capable of combining the virtues of democratic pastoralism with a progressive, energetic, and communal urbanity."[13]

THE NATURE OF PLACE

Many of us find expression for our relationship with the natural world in terms of places particularly meaningful to us. Our sense of place may be associated with a real place we have experienced or know indirectly or it may be found through the works of others in one of the many arts. In the latter instance place may be conveyed through persona, often in the form of latter-day archetypes arising from a vibrant collective unconscious.

Social scientists conducting formal analyses of landscape preferences anticipated differences among individuals, groups, and cultures. Instead they discovered comparable responses to natural scenes across those lines. People generally prefer the natural over the constructed or any other suggestion of human intrusion, let alone pollution. Urban scenes containing natural elements, notably trees and vegetation or water, are preferred over those lacking a suggestion of a natural component.[14]

Another approach to analyzing the meaning of place can be expressed in terms of a "genius" or "spirit" providing the qualities perceived there. Spirit in this context is intended in two senses: that of atmosphere or character and that of an "entity" with "its own soul and personality."[15]

THE REALMS OF PLACE

Cultural

For us the transition from mere physical location to place can be described as a process of experiencing deeply and place itself defined as a portion of the environment claimed by feelings: As a source sustaining our humanity the earth is a collection of places,[16] each unique to the individual collector.

This relationship can be traced at least into our European historical past. The sites and orientation of our ancient edifices were neither chosen at random nor constrained by formality or systematic reasoning. Intuition and symbolism played their parts. Today we might be said to be suffering a "geomantic amnesia" distancing us from such influences. As a result tourism may be a search for lost memories, a kind of *unconscious* pilgrimage, ostensibly secular in nature.[17]

Individually, our lives no longer depend on a close and deeply aware relationship to either the land or the sky. We rely on skilled experts to attend to purely objective omens and signs. Our personal disposition toward landscape is no longer a matter of life and death. And yet, personality lingers there; a people's perceptions "wash over the land like a flood, leaving ideas," although no one can tell the whole story.[18] There are moments when we can sense something of the story or even tell a portion of it. We feel connected somehow, if only momentarily.

Aesthetics determine the places we find extraordinarily beautiful and are the reason we set aside lands.[19] These are healthy beginnings to something deeper and promising for prospective national policy. A developing awareness of continuities in place can lead to the means of tending and restoration. An emotional connection can lead to survival.[20]

It is telling that contemporary architects and planners are beginning to see towns in a more natural light. One team asserts that towns should be designed for walking as well as driving and urges a return to towns made up of neighborhoods. To these planners it is not a matter of turning the clock back but of calling back and adapting features of the past that were successful. Despite some professional criticism at least one such town has been well received by the public as well as academics and professionals. Significantly, the express intent is to impart a sense of place and to think in holistic and relational terms.[21]

Children's

For a better grasp of our sense of place there can be no better beginning than our children's relationships to place. Play in biological infancy is generally held to be a form of pretense and practice for maturity. All animal young play, but only humans pretend to be other animals, a plant or even an artifact, such as a train. Children are in the process of synthesis of self and world. As Walt Whitman put it, a child may well become the first object he or she observes in the course of a day. That object becomes a part of the child for a time. A sense of place is a crucial component of the play and the process.[22]

Each child manifests an evolutionary biological heritage in the individual process of transcending nature. Each can be deemed a continuation of nature's own history who is capable of retaining a sense of individual identity, participating in surrounding "otherness," and maintaining an "ecological sense of continuity with nature" — an aesthetic and joyful "power to know and to be." In brief, childhood is "the point of intersection between biology and cosmology, where the structuring of our worldviews and our philosophies of human purpose takes place."[23]

The kinds of places children seek out are informative. The "nook" is a favorite, whether a tunnel behind a sofa, a space under a grand piano, or, better, a natural cave, a hollow in an overgrown bush, or a treehouse. Here, in that ubiquitous child-sized nook, the child observes the "beckoning openness beyond." "For the child nestled there, time comes almost to a stop: past and future cease to exist, displaced by a transcendent present."[24]

Memoirs frequently express this childhood penchant, which occurs across cultures and over time. Such memories are described as markedly similar moments of vision accompanied with a creative impetus upon recollection. The most striking of these recollections reflect "an acute pleasure in the incoming flux of minutiae," often a "mosaic of immediate sensory experience of the natural world" with "a sudden exultation and delighted sense of freedom in the vastness of open spaces." There may be a "momentary sense of discontinuity," an awareness of "unique separateness and identity" in company with a "revelatory sense of continuity," an immersion of one's being in the outer world. This apparently universal experience suggests a comparably "universal link between mind and nature . . . latent in consciousness."[25]

Personal

This place can be identified as one's intensely personal home and its loss by exile is devastating even, as transcendental poet Lord Byron observed, if the exile is "free to roam in the wide world." The exile, for whom the whole world is a dungeon, agonizes "in his vast prison-house" like a prisoner in a cell with bars of seas, mountains, and the horizon. Nature intensifies his sorrows by force of contrast.[26]

Today, our cherished places may not be immediately accessible or even real. Instead, we find them "in drama — in narrative, song, and performance." Significantly, this is an intensely personal albeit ubiquitous experience: "It is precisely what is *invisible* in the land . . . that makes what is merely empty space to one person a *place* to another." It is this place that "is suffused with memories or is a focus of stories, sacred and profane."[27]

Atavistic

Frequently, a sense of place is of the imagination, with no immediate correspondence to a real place. Each such experience can evoke a variety of emotions, feelings, or state of mind. The evocation may arise from the individual's own emotions, preoccupations, memories, and the underlying "biological, cultural, and religious heritage," all influenced by the place.[28] It comes out of our "entire evolution and the imprint of milennia" on our consciousness. "It is being in tune with waters and rocks, with vistas and horizons, with constellations and the infinity of time and space."[29] It can be largely atavistic.

This atavistic sense of wholeness can arise at the odd moment. While clamming more for fun than sustenance, Robert Finch pauses to contemplate the October sunset and ponders how it is that the play of light and color in sky and water drives "us into states of open-mouthed wonder and a feeling of sudden evisceration." The wondering pause may be deemed "useless, isolated, and distracting," but the fact remains that "nothing can move us more deeply than such moments of intense and unanticipated apprehension."[30]

For those of us who have been moved by this very sensation, wholeness accompanied by melancholia may be more accurate descriptions than evisceration or apprehension, but personal experience does call to mind a deep sense of something missing or lost, which is akin to a sensation of evisceration. A longing or pining for something lacking an identity is present.

A return to sixteenth-century German interactions among history, geography, and landscape art reveals forests as representative of atavistic wilderness paradoxically (and prophetically) serving as "places of health and wealth." It is here that the wild man image is said to have been converted from brute to noble savage, associated with a "better-groomed version of wildness." At this point wild men became "exemplars of the virtuous and natural life." Again prophetically this wild man "had to be wild enough to be distinguished from the effete . . . townsmen, but not so wild as to incur . . . accusations of brutishness."[31]

It is through the largely fictional experience of the wild man, the noble savage, the archetypal Wild Man that the balance between atavistic wilderness and civilization is often expressed for us, both linked strongly to a sense of place being articulated between author or artist and audience.

Spiritual

Even in the foregoing descriptions, the spiritual hovers about the picture of place. That realm may be more obvious to First Nations and to Eastern or primal cultures than to ours, but contemporary Christians express it. Popular novelist, sociologist, and Catholic priest Andrew Greeley mentions place directly and indirectly in his autobiography. He notes that people from Ireland find a sense of place in his novels. Greeley accepts this experience as an "equable marriage" of "geographical country" and that of the mind. The novelist himself acknowledges being "a confirmed 'local,' rooted in places and passionately loyal to them." He imposes on the reality his "own spiritual and emotional and intellectual country" and identifies his places as "sacraments, sacred and revelatory," with reference to "the strong needs of our fundamental nature as creatures constituted in time and space."[32]

Place, namely Ireland, is profoundly a part of one of Greeley's sermons. The moral is that, to enter heaven, the soul must let go of all worldly loves. Against the Lord's unbreakable rule one such Irish soul resists all manner of tricks in stubborn resolve to carry in a bit of Irish turf. At last the soul lets go of this final bit of earthly connection and enters heaven — only to discover that soul's particular heaven to be very much like Ireland.[33]

MEMOIRS OF REGION

Certain places throughout the regions of this nation have exerted their own influence inspiring eloquent responses. For those who love the New England shore there is Thoreau's description of Cape Cod's "long line of breakers rushing to the strand" on a day when the sea was dark and stormy and the sky completely overcast. A sympathy between wind and the sea is perceived. "The waves broke . . . as if over so many unseen dams, ten or twelve feet high, like a thousand waterfalls." The surf is also described as "droves of a thousand wild horses of Neptune," "with their white manes streaming far behind." Thoreau then muses that we do not associate "antiquity with the ocean, nor wonder how it looked a thousand years ago, as we do the land, for it was equally wild and unfathomable always." He concludes that the ocean is a wilderness, wilder even than a Bengal jungle, from which no city, however civilized, can "scare a shark from its wharves."[34]

A midwestern farm inspired Laura Ingalls Wilder to publish an article in 1923 in which she shares her childhood memory of bringing in the cows. She attributes the persistence of that particular memory to the peculiar attunement of a child's mind to nature's beauties, the "voices of the wildwood" and their deep impression. She laments that nature's voices cease speaking so plainly as we grow older and warns us that "in our busy lives, we neglect her until we grow out of sympathy. Our ears and eyes grow dull, and beauties are lost to us that we should still enjoy." Wilder urges us, no matter how much we enjoy our work, to "pause to watch the glory of a sunrise or a sunset" and promises that "a bird's song will set the steps to music all day long."[35]

The plains are said to "shape modes of thinking and feeling." There the vast lands, occasional butte, the permanent presence of the winds all combine in demanding a response requiring residents to go beyond the trivial. "It takes time and courage for people to find themselves at home in such an environment," but when they do they become "genuinely present in the environment."[36]

The far north is also evocative. Sigurd Olson, describing the north as a melody loud and clear, recalls becoming a part of it from childhood. He describes a hunger to live there and to know the place as fully as other denizens did instinctively. In later years that song still is heard with gladness. The reaction is more than a matter of terrain, forests, or lakes; it reflects "freedom and challenge of the wilderness," for which a longing was constant.[37]

Olson reportedly was moved by the silence of the north, described as "a primordial thing" seeping "into the deepest recesses of the mind, until mechanical intrusions were intolerable." There are other places and times when any sound seems a sacrilege. In such silences when natural sounds prevail, there comes "a deep awareness of ancient rhythms and the attunement" sought but seldom found. Such times and places must be compatible with others as well as evocative for the individual listener. Elsewhere Olson describes leisure as a kind of silence allowing one to become a part of creation, true leisure as companionship with the gods and expresses the anticipation of becoming "part of a golden, timeless world."[38]

Evocative place need not be remote wilderness. Jerry Mander speaks of a place in Yonkers where he walked to school through the woods. His recollection of details includes tangles of roots and a particular maple. He kept track of such changes as erosion following a rain. His mother dazzled the child when she advised him that was how the Grand Canyon started. (Instead Mander's path was converted into the New York Thruway.)[39]

A special place can remain urban. Terry Tempest Williams describes a colleague whose Bronx apartment overlooks Pelham Bay, the proximity aiding her sanity. Pelham Bay is home, "the landscape she naturally comprehends, a sanctity she holds inside."[40] In the end place is something from within. In a slim volume of minimal text matching a series of vivid place photographs, Joan Sauro describes something as seemingly mundane as a bus trip from her present home in Albany to that of her childhood in Syracuse. She speaks of a hush falling among the passengers as the road cuts through rock faces. Her own reaction of passing from place to place is one of "viewing my inner landscape for the first time. All the while the bus was going far and wide, I was going deep and deeper."[41]

Like so many others, William O. Douglas speaks of a special place, for him an island of "mostly rock and less than an acre in size," "the shoulder of a granite mass polished smooth by glaciers that slid slowly for centuries down from the north." That island is a "place of solitude bounded by blue waters where beavers splash, loons call, and ducks feed on wild rice." Douglas first saw his island place in moonlight and then in the "full flood-lights of the sun" with a "thick mist . . . rising from the waters." He describes it at night when a "chill settled over the waters. A thin mist started to rise at the foot of the falls. Moonbeams dance on waters. . . . A loon called somewhere in the darkness behind us."[42]

PLACE IN THE ARCHITECTURE OF FRANK LLOYD WRIGHT

The architecture of Frank Lloyd Wright is a form of memoir very much connected to place. In fact, this artist associated himself with the Celtic poet-prophet, Taliesin, in a shared "complex relationship with . . . levels of existence of which the physical world . . . is only one numinous manifes-tation." Wright's home dubbed Taliesin represents in terms of both site and structure "a Celtic reverence for the sacredness of place."[43]

The architect did not impose the building on the landscape but inserted in into a hillside, sensing "the effects to be gained from the contrast between a sophisticated . . . style and a gentle, pastoral setting," represent-ing his "vision of Arcadia, of . . . living in harmony with nature."[44]

A house, to Wright, must be natural, a "part of nature" and encompass-ing it, the two "seamlessly linked" in unity, part of a larger whole. His prairie houses stress the midwest's horizontal lines, with spreading roofs, bands of windows, and porches stretching into gardens, whereby the "house and its setting would merge and blur" to form a "single harmonious whole" with natural materials within as well as on the outside. Perhaps the most evocative of Wright's designs is Fallingwater in Pennsylvania, where the home is intricately united with its site, cantilevered over a ravine with its literally falling water.[45]

AVOCATIONAL EXPRESSIONS OF PLACE

Tourism and our other avocations can be expressions of place. Hunting is a prime example. Lopez identified hunting as a state of mind meaning "to have the land around you like clothing." This point leads to his definition of dreaming: "a nonrational, nonlinear comprehension of events in which slips in time and space are normal," equivalent to "the conscious working mind of an aboriginal hunter." We, unlike that aboriginal person, are irrevocably separated from the world of animals. That separation is likened to being separated "from light or water."[46] Today's recreational hunting is by inference a striving to breach that separation.

British royalty repair to the Scottish highlands, American honeymooners to the Poconos. An ancient fear and reverence was transformed in the nineteenth century "into havens for recreation and restoration" by the Romantic movement and intensive exploration of wild places, but it is clear a powerful symbolism lingers. It can be found in the sublime — Martin Luther King's thanksgiving for being allowed to go to the mountain — to the mundane preference for the upper floors of urban highrises.[47]

Similarly, the martial arts are asserted to be in some part about cosmology, a "transmission of 'deep knowledge'" symbolizing our place in the universe. We are starved for "any significant ritual," and the martial arts seek "an overarching theory of the universe" allowing the practitioner "to control the chaotic and perilous heart of life."[48]

The graphic arts have been another source of leisure activity. In the seventeenth and eighteenth centuries, the "Claude glass," named for a seventeenth-century French landscape artist, was carried into the countryside. When a suitable prospect was located, the user turned away and raised the glass for a kind of rearview image distorted by the glass's slightly concave and tinted facet.[49] Landscape art itself has been popular on both continents for at least as long. Contemporary versions include the black and white photography of Ansel Adams and the watercolors of Andrew Wyeth and, among aficionados of science fiction and fantasy, the fantastic land- and space-scapes of matte artists and the illustrators of book covers.[50] Within the obvious connections with the natural world "are hidden the dynamics of cultural continuity and nature attitudes." While influences among such works are of significance to art critics, for us the similarities "have a more profound significance as a mainroad of the evolution of the perception and imagery of nature."[51]

The juxtaposition of technology-bound civilization with something wilder or at least more evocative, made express in the name of George Lucas's special effects establishment, is particularly appropriate in the context of this chapter, this book, and future environmental stewardship. We need to combine our technological "wizardry" with our sense of wonder for both the material and the magical or spiritual realm.

The pastoral returns in this context. Its discovery by artists has been placed among the most important aesthetic events of our time. Going beyond the countryside's resources, nostalgia lent the pastoral an innocence and solace spanning centuries and remains a dominant theme today. At the same time wild nature, described as wild, beautiful, and sublime, continues as another prominent theme.[52] Legends and lore, the collective unconscious, archetypes, and their modern counterparts in the popular arts serve with legend to express that harmony with the universe or to mourn its loss.

NOTES

1. Stegner, "The Sense of Place" (1992), 42.
2. Frome, *Promised Land* (1985), 18.
3. Swimme and Berry, *The Universe Story* (1992), 127.
4. Oelschlaeger, *The Idea of Wilderness* (1991), 3–4.
5. Waterman, *Wilderness Ethics* (1993), 34, 41.
6. Stegner, "The Meaning of Wilderness for American Civilization," in Nash (1990), 176.
7. Machor, *Urban Ideals and the Symbolic Landscape of America* (1987), 6, 107.
8. Bruchac, *Survival This Way* (1987), 198.
9. Stilgoe, "Defining the Young Nation" (1995), 66–67.
10. Ibid., 15, 88.
11. Machor, *Urban Ideals*, 168–169.
12. Ibid., 88, 91.
13. Ibid., 91, 110.
14. Kellert and Wilson, *The Biophilia Hypothesis* (1993), 92–97.
15. Sheldrake, *The Rebirth of Nature* (1991), 173.
16. Devall and Sessions, *Deep Ecology: Living As If Nature Mattered* (1987) 111.
17. Sheldrake, *The Rebirth of Nature*, 178, 180, 181.

18. Lopez, *Arctic Dreams* (1986), 272–273.

19. Callicott, "The Wilderness Idea Revisited," in Chapple (1994), 170.

20. Kellert and Wilson, *The Biophilia Hypothesis*, 164.

21. Eric Morgenthaler, "Design for Living: Old-style Towns Where People Walk Have Modern Backers," *The Wall Street Journal*, February 1, 1993, A12.

22. Cobb, *The Ecology of Imagination in Childhood* (1977), 29, 31, 32.

23. Ibid., 18, 22, 23.

24. Tuan, *Passing Strange and Wonderful* (1993) 22.

25. Cobb, *The Ecology of Imagination in Childhood*, 87–88. See also, e.g., LaChappelle, *Sacred Land Sacred Sex* (1988), 118; Olson, *Reflections from the North Country* (1976), 112–114.

26. Jamil, *Transcendentalism in English Romantic Poetry* (1989), 183.

27. Lopez, *Arctic Dreams*, 278, citing American geographer Yi-Fu Tuan.

28. Sheldrake, *The Rebirth of Nature*, 164.

29. Mosher, *Songs of the North* (1976), 112.

30. Finch, *The Primal Place* (1983), 197.

31. Schama, *Landscape and Memory* (1995), 95.

32. Greeley, *Confessions of a Parish Priest* (1986), 83.

33. Ibid., 502–506.

34. Thoreau, *A Week on the Concord and Merrimack River — Walden; or, Life in the Woods — The Maine Woods — Cape Cod* (1985), 888–889, 980–981.

35. Hines, *Laura Ingalls Wilder* (1991), 119.

36. Jabner, "The Spiritual Landscape," in Dooling and Jordan-Smith (1989), 194.

37. Mosher, *Songs of the North*, 72, 70, 69.

38. Ibid., 133, 101–102.

39. Mander, *In the Absence of the Sacred* (1991), 12.

40. Williams, *An Unspoken Hunger* (1994), 39–48.

41. Sauro, *Whole Earth Meditation* (1986, 1992), 11, 13, 17.

42. Douglas, *My Wilderness* (1961), 99–103.

43. Secrest, *Frank Lloyd Wright* (1992), 9.

44. Ibid., 9.

45. Ibid., 113, 153, 425.

46. Lopez, *Arctic Dreams*, 200.

47. Gallagher, *The Power of Place* (1993), 72–73.

48. Donohue, *Warrior Dreams* (1994), 14, 15.

49. Chapple, *Ecological Prospects* (1994), 171.

50. See, e.g., David Kyle, *The Illustrated Book of Science Fiction Ideas & Dreams,* 1977, New York: The Hamlyn Publishing Group Ltd.; Michael Whelan, *Wonderworks: Science Fiction and Fantasy Art,* ed. Polly and Kelly Freas, 1979, Virginia Beach: A Starblaze Special; Charles Champlin, *George Lucas: The Creative Impulse,* 1992, New York: Harry N. Abrams, Inc.; J. M. Dillard, *Star Trek: "Where No One Has Gone Before": A History in Pictures,* 1994, New

York: Pocket Books; Thomas G. Smith, *Industrial Light & Magic: The Art of Special Effects,* 1986, New York: Del Rey.

51.   Shepard, *Man in the Landscape* (1967, 1991), plate 4, between 170 and 171.

52.   Ibid., plate 3, between 170–171, plate 11, facing 171.

IV

POPULAR CULTURE: OUTLETS AND SURROGATES

12

Place and Ambivalence in the Popular Arts

FROM THE SUBLIME TO THE RIDICULOUS?

There may be some suspicion that entry into American popular arts is a devolution from the sublime to the ridiculous. Nothing could be further from the truth, especially in terms of reaching the public at large in company with scholars and professionals. For the most part we are attuned to our culture more through the popular arts than through scholarly endeavors. As a replacement, if not an extension of earlier expressions, popular culture is a force to be reckoned with and acknowledged for its contributions.

There are those who assert that our young folks are not keeping up their respective traditions because they are passing too much time with television. Arguably, the resultant conviction is that the individual's "own visceral experience of the world hardly matters and that predigested images hold more truth and power than the simplest, time-tried oral tradition."[1] But in a nation where oral tradition has faded away, television is only one of numerous carriers of images and archetypes. After all, both Sophocles and Shakespeare wrote for their respective hoi polloi.

For the scholar, delving into the popular arts has its own reward. There the scholar "can be freer to be creative and playful" than is allowed in the more formal approaches. The scholar can be free of conventions and even shed some light on the subject of anthropology (for one) while communicating beyond the Ivory Tower. Our society accepts a broad spectrum of expression — including, for example, the martial arts — to fulfill psychic needs. Notably, the rewards of martial arts include a search for identity and reflect the generation of a coherent worldview and an investment of meaning into life. Thus, the martial arts, like so many popular arts, are a component of our "realm of myth and symbols."[2]

We are in no position to "insist that the adventure of self-discovery belongs only to the few who can carry it off with intellectual excellence in full public view." Whatever its form — Gaia, the environment, the cosmos — nature speaks to something within each individual "that wishes to be known peculiarly and personally, the self that fits no mold, the 'me' that cannot be made interchangeable." Certainly, we have no right to establish either genius or scholarship as prerequisite to self-knowledge.[3]

Fiction is one form that can merge with fact to form a "spirit of truth." Fiction fills gaps without documentation and without negating or reversing facts or necessarily engaging in mythologizing or outright propaganda.[4] Indeed, Jung noted that "[l]iterary products of highly dubious merit are often of the greatest interest to the psychologist." These are the works where the author has not engaged in "a psychological interpretation" of the characters, thus leaving the opportunity open to psychologist and reader, who may find an outlet for his or her own need for expression through identification with those characters. In the same literary breath, Jung mentions Herman Melville and H. Rider Haggard. Such works are grounded in "implicit psychological assumptions" of which the author may remain unconcious.[5]

Those writers who pursue the sentimental or romantic theme of the ideal rural home in natural surroundings have always been popular and influential. Most of their readers are urbanites unable "to pack up and desert the city"; such works are their "link to a more natural, civilizing order."[6]

As our distinctive Western genre lauds the outdoors, the lingering influences of European ideas, history, and fashions become increasingly remote. The characters of this genre are marked by "energetic individualism, great physical competence, stoicism, determination, restlessness, endurance, toughness, rebelliousness, resistance to control" — a residue of the Hero in the residue of wilderness.[7] In short, they are what many of us would be, were we able and in the proper setting.

A recurring theme in our popular culture is the tension between individual liberty and social constraint. "In the myth of the American warrior, this conflict is repeatedly celebrated but rarely resolved," perhaps because it remains alive within us as individuals. Of these warrior-heroes, the cowboy in particular exists "on the margins of respectability and economic utility . . . for a very good reason, for it is on the borderline that one finds both independence and danger."[8]

If there is any genre that speaks for us as individuals in search of something, environmental or other, it is science fiction. While hardly unique to America, science fiction and fantasy have a long history of response to their contemporaneous world. Utopian and satiric fiction responded to 1871's imminent Machine Age. Utopians were optimistic while satirists doubted. Another element in the emergent genre was increasing public awareness of the theory of evolution. On one side there was optimism in the ability of the machine to enable individuals to develop fully. On the other was the fear of enslavement by machines. Evolution was the mechanism of either prospective change.[9]

Utopia in the late nineteenth century features gardens and tillers of the soil, the pastoral image. Another theme of the time is decay of civilization through excessive use of machinery, with a consequent artificiality. Thereupon, humankind return to barbarism, with science fiction telling of "the Future-Past," a barbaric or atavistic Simple Life. Whatever destroys civilization, the survivors are reduced to savagery and rebuilding civilization.[10] A contemporary cinematic counterpart is found in the Mad Max trilogy and to a lesser extent in *Waterworld*.[11]

Whatever the themes, serious study of the genres within science fiction reveals without doubt that such romances have helped in the shaping of contemporary attitudes of all kinds of people, from college students to political leaders. It is that larger public whom these leaders must represent and reflect. Indeed, "the masses who read fiction for entertainment are inclined to be impatient of theory; the generalization, heavy with meaning for the philosopher, has little weight for . . . most people of the world."[12]

In its values, the genre can be compared to myth: "The modern Hercules, Thor, King Arthur, and Jack-the-Giant-Killer." This twentieth-century Hero, wherever he resides, "is Superman; in a series of world-wide best-sellers, he is Tarzan. His space-ships, death-rays, and invasions from Mars are Machine Age equivalents of Thunderbolts, magic carpets, and the sword Excalibur."[13]

Among the newest of the media, cinema comments "visually and metaphorically on the dreadful dilemma" constantly confronting us all: "how to construct our identity, how to get the image we offer to the world" at least nearly right. In film, the American hero may be urbane, but he is still a true man garbed in business suit ("true" suggesting something close to wild). Gardening is not viewed as low-paid manual labor but positions a "character as someone in tune with the outdoor life," "healthy and 'male.'" Forestry suggests strong individuality, associations with the earth, rootedness and solidarity. Atavism manifested in a variety of persona is extremely popular.[14]

The warrior myth, from the Western hero to Tarzan to Superman to Rambo, celebrates on our behalf "the necessity of the warrior while upholding the primacy of the group." Thus, our continuing "fascination with the mythic warrior" is integral to "our social dynamic" and fundamental to "the American cultural blueprint . . . celebrated and hence reinforced with each successive telling."[15]

IMAGES OF PLACE

The Roles of Nature in Place

Along with the Simple Life, place is an important element in popular culture. The child's places are loaded with "memories, myths, and meanings"; the adult's "more elaborately wrought." In either case, our propensity to separate nature from human perception is to ignore something indivisible. Landscape, in fact, "is built up as much from strata of memory as from layers of rock."[16]

Withdrawal to a remote place seems to be a variation on an American dream. Robert Finch speaks of looking up from typing on a May morning to view "the stiff, gray, intersecting patterns of the oak trunks and branches" beyond his window. It was like seeing their reality for the first time, "a vast, living maze stretching out beyond my lines of sight." Suddenly the author knew beyond doubt that what he wanted in this place "was entrance, or rather re-entrance, into the maze." He speaks of his preference "to go silently and smoothly into the maze, without a rustle, as the light fox bounds with inborn agility across the rounded stone wall" and as other denizens make their way through their natural homes.[17]

Jim Brandenburg experiences the forest in much the same way, with reference to childhood entry. Game trails lead as the forest beckons to draw the child in where he stops and looks and touches and imagines until something extraordinary occurs. "Something was out there. Something mysterious and powerful. Something old. Part of me wanted to go deeper and deeper." As an adult he calls the wilds of Minnesota forest his cathedral, dream, and sanity.[18]

The reaction to place can be extraordinarily personal as is suggested in a response to old photographs of a place once experienced, even when the precise location cannot be recalled. Lack of precision does not preclude one from going back to that place. In fact, *because* the meadow lacks fixed coordinates it is possible "to return to it again and again." First encounters with nature are described as immersions. "The particular aspects of a place were less vivid than the overall sensation of being surrounded by wildness." And being liberated from particulars, the place in the photographs can stand in for all places evoking the initial "oceanic sensation." Photographs become windows "onto a private Eden, a place for all time," even though one "can no longer reside there." "[U]nder the right circumstances it is possible to visit, if only briefly."[19]

Place in Ayn Rand's *The Fountainhead*

Place can insinuate itself into the most unexpected contexts. For those who are familiar with the works and philosophy of Ayn Rand, that she might express this sense of place comes as an illuminating surprise.

Roark, architect-protagonist of *The Fountainhead,* passes some time working in a quarry. He makes a habit of walking into the woods and stretches supine on the ground to "watch the patterns of veins on the green blades of grass under his face." After studying them for a bit, Roark "would roll over on his back and lie still, feeling the warmth of the earth under him." Characteristically, he exacts a small victory from the resistant earth, but it is with a "dim, sensuous pleasure" suggesting a kind of connection.[20]

Roark's foil, Keating, is a frustrated artist who ultimately rents a shack in the hills where he sets up an easel. From that vantage "he looked at an empty sweep of hills, at the woods and the sky, He had a quiet pain as sole conception of what he wanted to express, a humble, unbearable tenderness for the sight of the earth around him — and something tight, paralyzed, as sole means to express it. . . . There was no pleasure in it, no pride, no solution; only — while he sat alone before the easel — a sense of peace."[21]

There can be no doubt that Roark's architecture is inspired by and closely linked to place. Even among his earliest sketches are personal exercises done "when he found some particular site and stopped before it to think of what building it should bear." He observes to the dean of the school expelling him: "No two materials are alike. No two sites on earth are alike. No two buildings have the same purpose. The purpose, the site, the material determine the shape. Nothing can be reasonable or beautiful unless it's made by one central idea, and the idea sets every detail. A building is alive, . . . [i]ts integrity is to follow its own truth, its single theme, and to serve its own single purpose."[22]

Why is Roark both the architect and maverick he is? "Because I love this earth. That's all I love." And he expresses that unlikely love in terms of the "honest residence" he longs to design in Connecticut on "a lonely, rocky stretch of shore, three miles beyond an unfashionable little town; . . . a cliff rising in broken ledges from the ground to end in a straight, brutal naked drop over the sea, a vertical shaft of rock forming a cross with the long, pale horizontal of the sea."[23] Roark was to return to this place numerous times. The ultimate design is not his but that of the cliff: "It was as if the cliff had grown and completed itself and proclaimed the purpose for which it had been waiting." Levels followed the ledges, "in planes flowing together up into one consummate harmony," with walls of the same granite carried upward and terraces "silver as the sea" following "the line of the waves."[24]

A secondary character is in tune with place in the same senses. As he is biking along a wooded Pennsylvania trail in a "beauty almost tangible" and feeling "the fresh wonder of an untouched world," he is pondering the unnamed thing he wants of life, something he feels "in this wild loneliness." He experiences anger because his exultation is experienced only in the wilderness and is lost when he returns to civilization. He is looking to music or elsewhere for the genius.[25]

And then a valley opens up in front of him to reveal a town:

There were small houses on the ledges . . . before him, flowing down to the bottom. He knew that the ledges had not been touched, that no artifice had altered the unplanned beauty of the graded steps. . . . [T]he houses [had become] inevitable, and one could no longer imagine the hills as beautiful without them — as if the centuries and the series of chances that produced those ledges in the struggle of great blind forces had waited for their final expression.[26]

If Ayn Rand — sophisticated, urbane, unabashed proponent of humanity's civilized devices — can, through her characters, demonstrate so clearly a deep appreciation for place, there is certainly hope that each of us can find a surrogate if not an outlet for our own relationship to the world around us.

REFLECTIONS ON THE SIMPLE LIFE

Atavism

Probably the most evocative reflection of the Simple Life is atavism. Looking back on her childhood, Clarissa Pinkola Estés distinguishes wild from human time in the context of remembering dozens of seasons, describing some of them as night thunderstorms, heat lightning, forest bonfires, blood-on-the-snow, and ice, blowing, crying, shimmering trees.

She speaks of loving the seasons of snow: diamond, steaming, squeaking, dirty, stone, the latter two promising the coming of "the time of flower blossoms on the river."[27]

Children retain their wild nature where once our lifelong habit was to live by cycles and seasons as "they lived in us. They calmed us, danced us, shook us, reassured us, made us learn creaturally." In today's world a periodic return to "the wildish state" replenishes our psychic reserves.[28]

A tinge of the atavistic can lead to a melancholia beyond poignancy, as is described by one on a walk reemerging from an abandoned residential development to find the remains of a campfire in the woods. "The fire had a poignant forlorn look about it, as though it might have belonged to the ghost of some old Indian who had climbed up out of the swamp mists, taken a brief, sad glance around, and then disappeared down the embankment again."[29]

The noble savage remains a viable, evocative image. It is an idea of width and magnitude, still representing "a new ideal of human political relations" after more than five hundred years.[30] The genres including a variety of noble savages are very much alive today. Popular authors still believe in the value of living near to nature without the artificiality of civilization, but so must readers or the genre would die out. Where England is the stand-in for civilization, English noble savages can reconcile faith in English superiority with condemnation of urban superficiality. Tarzan of the Apes and his brethren "combine the splendid physical qualities of the jungle native with the intellectual and moral superiority of an Anglo-Saxon heritage,"[31] another manifestation of the ambivalence in our relationship with civilization and the wild.

For many who know Tarzan through his cinematic incarnations, his treehouse is an important element in the overall image of his tempered wildness. Treehouses themselves are an American icon not limited to childhood and may actually be universal in their appeal, perhaps as an escape "from an earthbound perspective." One answer to the question of why treehouses cause one "to pause, take a closer look and smile" may be an inborn affection lingering from the time "our ancestors lived in trees" and arising from "some primal memory" suggesting they are still our home.[32]

Some of these genre entries are appealing in their suggestion of a closer connection among peoples. In a council of peoples in one of those atavistic Future-Histories, "the genetic bequests of the ancestors" are "reflected in the skin colors and bone structures and textures of hair" of the participants. "[A]ll the waters of the world had flowed over this spirit-haunted land, leaving something washed ashore in their wake. Ivory, sepia, raw umber, burnt umber, ebony, charcoal, sienna — a palette of earth tones, like colors out of a paint box. Auspicious, they called it, when children of ancestors from all four directions sat together and the circle was whole."[33]

THE RETURN OF WILD MAN

The Wild Man Within

The psychological id in which the inner Wild Man is said to reside is also known as the conservative component of the psyche. Here also reside patterns of environmental adaptations, a treasury of ecological intelligence. This ubiquitous Wild Man resists those social forms endangering "the harmony of the human and the natural"; its untamed "selfishness represents a bond between psyche and cosmos" with origins reaching back to the Big Bang.[34]

Theodore Roszak urges us to reconnect with both our biological and our ecological selves and notes that at least one Freudian psychiatrist holds the belief that a child's mind "recapitulates the psychic phylogeny of life on Earth" before becoming the integral, independent entity that is the mature person. At least one Romantic poet suggests "that the acquisition of speech marks the end of our animistic sensibility"; thereafter "the child grows away from inherent intimacy with nature."[35]

It is the collective unconscious that shelters our "compacted ecological intelligence." The id and ego must be reunited, not only for the sake of sanity but also to allow "greater evolutionary adventures."[36] The fictional wild man's need to reconcile his wildness with his civilization is a metaphor for that human need.

A feral or wild child can be a spokesperson for a vitalist philosophy, where the adult is bound to fact with a loss of connection with the "flow of life," in some part because children are free of both guile and rationalist prejudice and able to "enjoy a genuine direct connection with" life.[37]

Ambivalence in Popular Fiction

Mason's Broken Road

This turn of the century genre novel is of an Indian prince, Shere Ali, and his friendship for an Englishman and love for an Englishwoman in a time when the prince's was deemed a subjugated, inferior race. Ali finds himself torn between the cultures of remote India and England. One summer evening just prior to his planned departure from England, the prince is in a conservatory "impressing upon his memory every detail of the place, the colours of the flowers and their very perfumes." He is listening to music, leading to a note of regret rising "in his ears" to dominate the melody. "Tomorrow the lights, the delicate frocks, the laughing voices and bright eyes would be gone." Shortly Ali would be on his way back to India, and he asks where he belongs. "On one side was all that during his ten years he had . . . learned to love and enjoy; on the other side was his race and the land of his birth."[38]

Later, in India, Ali encounters the woman he loves, who, although she does not return the sentiment, is obviously intrigued: "She wondered in what guise he would come to her: a picturesque figure with a turban of some delicate shade upon his head and pearls about his throat, or — as she wondered, a young man in the evening dress of an Englishman . . . came towards her."[39] It is Ali.

When urged to return to his home to marry among his own people, Ali can do neither, asserting that he had been harmed by those who sent him to England, because neither place can be his true home: "I am a citizen of no country. I have no place anywhere at all." Significantly, Ali meets an Englishman who is also unable to return home to England. Without any regrets, this man returned to spend the remainder of his life in India. Ali, however, must remain, and he cannot use his expanded horizons.[40]

Ali's wildness is aroused by circumstances. Angered by an Englishman, "there was very little of the civilised man in Shere Ali's look at this moment. His own people were claiming him. It was one of the keen grim tribesmen of the hills who challenged" the other. Later, Ali is described as the victim of the follies and illusions of chivalry, who "ached like a man bruised and beaten; he was possessed with a sense of loneliness, poignant as pain." Still later, he comes to hate England and the English and recovers his own place, "[t]he longing at his heart . . . for his own country — for his own people." And he rejoices at finding "his place in the world."[41]

Eventually, Ali leads his tribesmen in insurrection only to lose to the tactics of his former English friend. The Englishman pursues Ali over a span of years to find him essentially alone in the wild. Ali is brought before the English who banish him to Burma where he is drinking himself to death as the book closes.[42] Obviously, torn between two places, Shere Ali has lost both of them.

In Ayn Rand

Returning again to Ayn Rand's *The Fountainhead,* there is no more urbane sophisticate than Gail, the publisher. On his yacht he proclaims to the heroine, Dominique, that he finds no rapture from nature, but from skyscrapers. For one sight of the New York skyline Gail asserts he "would give the greatest sunset in the world." He further asserts he never feels small before the wonders of nature, but feels "the greatness of man" when he looks at the ocean. Thereupon Gail becomes rhapsodic over buildings, *because they fit their place and extend it.*[43]

The Literary Lineage

The modern lineage of the noble savage can commence with Kipling's Mowgli, believed by some to be the most direct ancestor of Tarzan. A reading of the two chronicles, however, strongly suggests Tarzan to be the more evocative of the two, especially in terms of a surrogate for balancing wildness with civilization.

The Jungle Books, including the tales of Mowgli, were written in the years 1894 and 1895. There is some suggestion they were influenced by *Uncle Remus.*[44] At least one, "Mowgli's Brothers," is told from the point of view of the wolves, Mowgli's adoptive family. These and all the other creatures of the jungle speak among themselves in ordinary English. All the creatures instinctively recognize a special destiny for Mowgli as man's cub.[45]

In fact, despite his relationship with the wolves and other animals, Mowgli maintains a certain superiority, manifest in a humorous moment arising from an encounter with the boy's enemy, a pack of dogs: "Mowgli stretched down one naked leg and wriggled his bare toes just above the leader's head. That was . . . more than enough to wake the pack to stupid rage. Those who have hair between their toes do not care to be reminded of it."[46] In humor there is one reference suggesting something of a dual nature, as Mowgli greets one of the creatures with a "Good hunting," thus carrying "his manners with his knife, and that never left him."[47]

By his teens Mowgli is a more nearly the noble savage of popular culture: "[H]e stood . . . strong, tall, and beautiful, his long black hair sweeping over his shoulders, the knife swinging at his neck, and his head crowned with a wreath of white jasmine."[48]

Mowgli is endowed with extraordinary physical abilities. At seventeen he is possessed of strength and growth far beyond his age. "He could stop a young buck in mid-gallop and throw him sideways by the head. . . . The Jungle-People, who used to fear him for his wits, feared him now for his mere strength. . . . And yet the look in his eyes was always gentle. Even when he fought his eyes never glazed as Bagheera's did. They only grew more and more interested and excited."[49] Mowgli's sense of smell is highly developed by human standards but falls short of the ability of a "jungle nose" and is further vulnerable to Mowgli's sojourns among villagers.

Mowgli's character is as much molded by his feral existence as is his physical being. He is human enough to want to lord it over the animals and to possess a sense of adventure. He flouts death as the animals must but with the uniquely human knowledge that he is doing so. While more human than animal, Mowgli has the characteristic ambivalence of the feral child. His fatalism is coupled with common sense and practicality. Like the best of both human and animal, Mowgli is a staunch ally.[50]

There are rare moments of tranquility in Mowgli's life. Some of these moments, even with Akela, Mowgli's foster father, are fraught with danger, both immediate and prophetic. When, for example, Mowgli swiftly strikes at Akela with his knife, both the circumstances and the thinly veiled threats of the two are telling:

Akela was a wolf, and even a dog, who is very far removed from the wild wolf, his ancestor, can be waked out of a deep sleep by a cartwheel touching his flank, and can spring away unharmed before that wheel comes on.

"Another time," Mowgli said, quietly, . . . "speak of the man pack and of Mowgli in *two* breaths — not one."[51]

Eventually Mowgli's human destiny comes to pass. The boy orders the pack away from its attack on a man and is chastised by Akela with reference to the boy's earlier response. As Bagheera observes, Mowgli's retort is all human, the inferiority of the beast is implicit. Mowgli is the superior, not by reason of his feral upbringing, but because he is human.[52]

The point is reinforced, despite Mowgli's reluctance to accept his difference, in one brief passage: "Mowgli looked at [Bagheera] lazily . . . and . . . the panther's head dropped. Bagheera knew his master."[53] Bagheera puts it most effectively: "'Thou art of the jungle and *not* of the jungle,' he said at last."[54]

With his dying words Akela warns Mowgli,

"Thou art a man, . . . wolfling of my watching. . . . Go to thine own people."
"I will never go. I will hunt alone in the jungle. I have said it."

When Mowgli asks who will drive him, the answer is himself, and Akela exhorts him to go to his people, to men.[55]

Torn between his life as a beast of the jungle and his undeniable human side, at first Mowgli clings to the former as any child clings to home and hearth. But the call of humankind is compelling.[56] His initial efforts to become human are less than successful,[57] and then he observes that he will progress until — "At the end I shall be Mowgli the man."[58]

His profound ambivalence actually renders him ill: "By night and by day I hear a double step on my trail. When I turn my head it is as though one had hidden himself from me that instant. I go to look behind the trees and he is not there. . . . I lie down, but I do not rest. . . . The kill sickens me, but I have no heart to fight except I kill. . . . I know not what I know."[59]

Sadly, Kipling never takes Mowgli into manhood. We will never know the feral man through Mowgli. For that we must proceed to Tarzan of the Apes.

NOTES

1. Nabhan, "Going Truant," in Nabhan (1994), 106.
2. Donohue, *Warrior Dreams* (1994), 1–3, 5.
3. Roszak, *The Voice of the Earth* (1992), 277–278.
4. Churchill, *Fantasies of the Master Race* (1992), 79.
5. Jung, *Modern Man in Search of a Soul* (1933), 154. Jung's point is one reason for the popularity of an *Ishmael* (Daniel Quinn; 1992, New York: A Bantam/Turner Book) or a *The Celestine Prophecy* (James Redfield; 1993, New York: Warner Books) and, simultaneously, an explanation for the sense of detached emptiness both convey.
6. Machor, *Urban Ideals and the Symbolic Landscape of America* (1987), 154.
7. Stegner, "The Meaning of Wilderness for American Civilization," in Nash (1990), 113–114.
8. Donohue, *Warrior Dreams*, 56, 58–59.
9. Bailey, *Pilgrims Through Space and Time* (1947), 51, 54.
10. Ibid., 55, 88, 255, 313.
11. *Mad Max* (1979), *The Road Warrior* (1981), *Mad Max Beyond Thunderdome* (1985); all directed by George Miller. *Waterworld* (1995), Kevin Reynolds, director.
12. Bailey, *Pilgrims*, 294.
13. Ibid., 293–294, 295, 306–315, 317–318.
14. Kirkham and Thumin, *You Tarzan* (1993), 18, 20–21, 29–44.
15. Donohue, *Warrior Dreams*, 66.
16. Schama, *Landscape and Memory* (1995), 6–7.
17. Finch, *The Primal Place* (1983), 3, 8.
18. Brandenburg, *Brother Wolf* (1993), 13, 11.
19. Dobb, "Back to the Future" (1995), 80–81.
20. Rand, *The Fountainhead* (1943, 1971), 203.
21. Ibid., 565.

22. Ibid., 19, 24.
23. Ibid., 49, 124.
24. Ibid., 124–125.
25. Ibid., 504–505.
26. Ibid., 505.
27. Estés, *Women Who Run with the Wolves* (1992), 56.
28. Ibid., 256–257. See also Torgovnick, *Gone Primitive: Savage Instincts, Modern Lives* (1990).
29. Finch, *Primal Place*, 73–75.
30. Weatherford, *Indian Givers* (1988), 130.
31. Street, *The Savage in Literature* (1975), 8.
32. Nelson, *Treehouses* (1994), 5–6.
33. Starhawk, *The Fifth Sacred Thing* (1993), 47.
34. Roszak, *Voice of the Earth*, 290.
35. Ibid., 295–296, 298.
36. Ibid., 304, 305.
37. Hermundsgård, *Child of the Earth* (1989), 51.
38. Mason, *The Broken Road* (1907), 111.
39. Ibid., 146.
40. Ibid., 156, 174.
41. Ibid., 237, 240, 261.
42. Ibid., x.
43. Rand, *The Fountainhead*, 447.
44. Kipling, *The Jungle Books* (1961).
45. Ibid., 11–29.
46. Ibid., 293.
47. Ibid., 242.
48. Ibid., 216.
49. Ibid., 304.
50. Ibid., 279–302.
51. Ibid., 189.
52. Ibid., 190.
53. Ibid., 305.
54. Ibid., 202.
55. Ibid., 301.
56. Ibid., 188, 196.
57. Ibid., 59.
58. Ibid., 291–292.
59. Ibid., 321.

13

Edgar Rice Burroughs's Tarzan of the Apes: Timeless Surrogate

While Tarzan of the Apes needs no introduction to those who know him through the writings of Edgar Rice Burroughs, the depth of this character will come as a surprise to any who know only the cinematic lineage epitomized by Johnny Weissmuller. In either version, the ape-man is a worthy surrogate for those who seek expression for their need to find their own unique balance between civilization and the wild. As the Wild Man, Tarzan has a wealth of kin in our popular culture.

Burroughs's Tarzan is, among other things, the classical hero described in Chapter 7. A reading of *Tarzan of the Apes, Return of Tarzan, Jungle Tales of Tarzan, Son of Tarzan,* and either *Tarzan and the Jewels of Opar* or *Tarzan the Untamed* is sufficient to reveal this Hero.[1] To the extent he does not actually fit every component of the pattern, modern parallels to the ancient themes can be discerned, although some may be out of chronological order or otherwise altered. Who, then, is this Hero and Wild Man, Tarzan of the Apes? What is the nature of Burroughs's unique contribution to literature and popular culture?

EVOLUTION

Background

Tarzan's parents are John and Alice Rutherford Clayton, England's Lord and Lady Greystoke. Lady Alice is pregnant when a mutiny aboard the ship carrying the couple to her husband's mission in Africa maroons the ill-fated couple on the coast. Their son is born in 1888. Lady Alice survives no more than another year, and Sir John does not long survive her.[2]

The child, also John Clayton, is adopted by Kala, a Grey Ape. By the time Kala's "Tarzan" is ten, he has the strength of a man in his prime, but he is far more agile than the most practiced athlete. Tarzan spends his first twenty years in the jungle.[3] He succeeds in teaching himself to read and print in English, possibly because the Grey Apes are capable of something very close to language.[4] The boy eventually rises to leadership among the apes. He is in his late teens before he encounters Caucasians. Tarzan is returned to civilization by the Frenchman, Paul d'Arnot, and eventually marries an American, Jane Porter.[5]

Whatever the critical response of his own time, Burroughs is one of a very few authors of any genre who can pick up a reader and hurtle him or her through each and every book without ever permitting a pause to ponder such niceties as style. Burroughs's works are thriving today while most of his contemporaries are long out of print. Both Burroughs and Tarzan stand virtually alone and unchallenged in their respective niches. The two survive because they grip their audience. And Tarzan is gripping because, while, like us, he is striving to find his place between wildness and civilization, he is adept in both settings although he is never completely satisfied in either. Tarzan, like us, needs elements of both civilization and wildness for his life to be complete.

In keeping with the distance he often maintains from people, Tarzan's sense of humor, like his rare smile, is selective. Unless associated with anger it tends to be suppressed except in the company of family and friends. At such times it is rather gentle, if a little foreboding. When dealing with civilization Burroughs is both broader and less tolerant than is the ape-man.[6]

Usually Tarzan's self-imposed isolation from the effete ways of civilization is enthusiastically applauded by Burroughs. It can, however, have its minor drawbacks: "Among the numerous refinements of civilization that Tarzan had failed to acquire was that of profanity, and possibly it is to be regretted since there are circumstances under which it is at least a relief to pent emotion."[7]

Physical Attributes

The popular physical image of Tarzan does not fall too far from the mark. At eighteen he is six feet tall. He is fearless with the stoicism of the apes. He has complete self-confidence and is mentally quick — far beyond the powers of the Grey Apes. Tarzan is repeatedly described as godlike, straight as a young Indian. While all his senses are developed to a preternatural degree, his sense of taste is least developed.[8]

As a mature man Tarzan is still described as godlike, albeit primitive: "Tall, magnificently proportioned, muscled more like Apollo than like Hercules, garbed only in a narrow G string of lion skin . . . , he presented a splendid figure of primitive manhood that suggested more, perhaps, the demigod of the forest than it did man."[9] That Tarzan has the fluid grace of the Olympian is reinforced time and time again, along with his animal nature: "The easy, majestic grace of his carriage; his tread, at once silent and bold; his flowing muscles; the dignity of his mien; all suggested the leonine, as though he were, indeed a personification of Numa, the lion."[10]

The ape-man's face reflects the best of both sides of his unique character: "A perfect type of the strongly masculine, unmarred by dissipation, or brutal or degrading passions. For, though Tarzan of the Apes was a killer of men and of beasts, he killed as the hunter kills, dispassionately."[11] His eyes, in particular, reflect that strength of character: "A shock of black hair falls in disorder above serene, grey eyes, eyes that can reflect the light of a summer sea or the flashing steel of a rapier."[12]

What renders Burroughs's Tarzan at once unique and fascinating is the combination of character and abilities derived from his two utterly different environments. This combination appears while Tarzan is still in his chronological youth and wholly naive to the ways of humankind:

He was fully grown now, with the grace of a Greek god and the thews of a bull, and, by all the tenets of apedom, should have been sullen, morose, and brooding; but he was not. His spirits seemed not to age at all — he was still a playful child, much to the discomfiture of his fellow-apes.[13]

Tarzan must truly be as intelligent as he is physically agile in order to survive: From infancy death had stalked, grim and terrible, at his heels. He knew little of any other existence. To cope with danger was his life and he lived his life as simply and as naturally as you live yours. . . . Tarzan had lived as the lion lives and the panther and the elephant and the ape — a true jungle creature dependent solely upon his prowess and his wits, playing a lone hand against creation.[14]

Like his fellow animals Tarzan is blessed with a sixth sense supplemental to his physical prowess. Usually this preternatural sense serves as additional protection as when a great cat is stalking him and the ape-man has no physical indication of his presence.[15] It is not always enough: "The abnormal sensibility of the hunted beast warned him of impending danger; but he did not know where to look for it, nor in what form to expect it."[16]

In every such passage, Burroughs brings the reader into identity with Tarzan while emphasizing the ape-man's feral nature and connection with the world in which he thrives from childhood into his later years. These affective passages continue through the whole series into the posthumous publications of the 1960s. Tarzan in Burroughs's hands never loses that wildness that at least borders on the Wild Man archetype and never strays from being the noble savage, as well as hero in the popular and archetypal senses. These qualities extend deeply into Burroughs's characterization of his creation.

Characterization

Intelligence

Dignity requires intelligence, with which Tarzan is amply endowed. From the beginning Tarzan's very survival is attributable to his quick mind. Moreover, in ways left unreported by Burroughs, the ape-man accumulates rather a wide-ranging education.[17] In point of fact, a reading of the entire series reveals this ape-man as something of a displaced Renaissance Man.

Nobility and Honor

Nobility and honor are inherent to Tarzan notwithstanding his bestiality. One of the most striking examples of his basic goodness occurs when his arch-enemy Rokoff attempts to compromise Tarzan with a young matron of the nobility. The ape-man is only recently removed from those twenty years in the jungle, yet his ultimate response to the lady and to her enraged husband goes far beyond being merely proper. As Burroughs observes, Tarzan and d'Arnot, his mentor in things civilized, share the same exalted ideals of manhood, personal courage, and honor.[18]

Tarzan is never at a loss as to what his own nobility demands, for all its naiveté. And yet he is never allowed to seem either pompous or foolish.[19] But always his duality prevents Tarzan from ever becoming too perfect to be an interesting study. All this heroic virtue is skillfully juxtaposed against his basically savage animal responses.

Animal Stoicism

One of the characteristics "making Tarzan Tarzan" and so appealing a role model is his animal stoicism: "A jungle beast was Tarzan with the stoicism of the beast and the intelligence of man."[20] When captured, the merged persona are particularly apparent: "[H]e did not reply, but only stared straight at [his captor] with cold and level gaze. The hyenas crept up behind him. He heard them growl; but he did not even turn his head. He was a beast with a man's brain. The beast in him refused to show fear in the face of a death which the man-mind already admitted to be inevitable."[21] This extreme of stoicism, though in part a matter of calculation, is not entirely human. It is one of Tarzan's most bestial traits and yet it is also strangely ennobling.[22]

Savagery

For the most part his savagery is a neutral aspect of Tarzan's character — if only in terms of himself, as is revealed when he watches mutineers bury the treasure chest of Jane Porter's father.[23] Savagery is both basic and very natural in his persona: "Tarzan . . . was ranging a district that was new to him, and with the keen alertness of the wild creature he was alive to all that was strange or unusual. Upon the range of his knowledge depended his ability to cope with the emergencies of an unaccustomed environment. . . . [I]n certain matters concerning the haunts and habits of game both large and small, he knew quite as much if not more than many creatures that had been born here."[24] He has the hunted beast's respect for firearms and an aversion for entering unfamiliar enclosures. When Tarzan is seriously ill or injured, his savage nature moves him to go into retreat — "He saw a tangle of almost inpenetrable thicket, and . . . he crawled into it to die alone and unseen, safe from the attacks of predatory carnivora."[25] What is primarily neutral does blend almost imperceptibly into the positive. Burroughs frequently reiterates that, while the ape-man is by necessity a skilled hunter and a killer, it is the natural skill of the beast of prey. He never kills for "sport."[26]

Significantly, when Tarzan is forcibly returned to his jungle habitat, after what appears to have been an absence of months, at the first threat of danger —

Instantly his senses awoke, and he was again Tarzan of the Apes.

As he wheeled about, it was a beast at bay, vibrant with the instinct of self-preservation that faced a huge bull-ape that was already charging down upon him.

Once again he was the jungle beast reveling in bloody conflict with his kind. Once again he was, Tarzan, son of Kala the she-ape.[27]

Making the character still more interesting, the positive is not entirely without its negative side. Again the blending can be subtle. When provoked, Tarzan becomes more savage than angry in the usual human sense, and he is the more dangerous for it. And his savage nature can emerge absent any need to protect self, mate, or young. At one point he methodically stalks another who has stolen diamonds from him.[28]

When he is first introduced to civilization, Tarzan's savage nature causes him to overreact and very nearly lands him a lengthy term in prison. Tarzan is at his ugliest (and Burroughs at his most controversial) when he commences to avenge Jane during World War I.[29] Fortunately, over the series, Tarzan's anger, with or without the savage touch, is sparingly aroused and serves the man well. It is controlled so long as that hovering savagery remains subdued. While anger can provoke him to revenge, it does serve to sustain the ape-man when most needed, as when he is in search of those who have in fact kidnapped Jane:

[I]t was a savage Tarzan who threaded the mazes of the soggy jungle. . . . Even the panthers and the lions let the growling Tarmangani [Caucasian] pass unmolested.

When the sun shone again upon the second day and a wide, open plain let the full heat of Kudu [the sun] flood the chilled, brown body, Tarzan's spirits rose; but it was still a sullen, surly brute that moved steadily onward.[30]

In keeping with his animal nature, Tarzan loathes captivity.[31] His need for freedom — the jungle — is found in the first and the last of the series.

Tarzan had no sooner entered the jungle than he took to the trees, and it was with a feeling of exultant freedom that he swung once more through the forest branches.

This was the life! ah, how he loved it! Civilization held nothing like this in its narrow and circumscribed sphere, hemmed in by restriction and conventionalities. Even clothes were a hinderance and a nuisance.

At last he was free. He had not realized what a prisoner he had been.[32]

Duality

While Tarzan's fine attributes are completely dependable, he himself appears to exert little control over which aspect of his self will come to the fore. When attacked by a bull-ape: "Instantly Tarzan was transformed from a good-natured, teasing youth into a snarling savage beast. . . . He did not wait for the bull to reach him, for something in the appearance or the voice of the attacker aroused within the ape-man a feeling of belligerent antagonism that would not be denied. With a scream that carried no human note, Tarzan leaped straight at the throat of the attacker."[33]

On those few occasions when the patience of the most loyal friend might be tried by his excessive calm, Tarzan's own response tempers the apparent hubris: "'I know Africa, and I know myself,' replied the ape-man, simply. There was no bravado in his tone, but absolute assurance."[34] This short passage sums up the whole man but fails to reveal the profound split in his personality, the characteristic that renders Tarzan so effective a surrogate for our own personal split between wildness and civilization.

Tarzan is, after all, feral man with all the strangeness and the deep ambivalence inherent in that concept. All of his traits are touched by the ambivalence — usually, but not always, for the better. He has the dignity both of the beast and of the heritage behind noble blood. It is a dignity that does not yield to pressure. When Tarzan declines to kneel before the queen of one of Burroughs's numerous lost cities, the ape-man explains to a friend that it is not a matter of foolish bravado but something within his being. It is something accepted by royal personages.[35] It remains intact when the man is a captive.[36]

The Lighter Side

Tarzan does have a lighter side, which appears early in his life in the jungle. This side of the ape-man is reflected in his smiles and laughter, both extremely rare to him. Obviously, the ape-man is more relaxed and, therefore, more inclined to smile when he is among his animal friends, notably Tantor, the elephant, and Nkima, the small monkey. Tarzan does smile at his own self-styled weaknesses and at the irony of circumstances — with a certain wry self-deprecation.[37] Laughter is by far more rare than the ape-man's smiles. Until his first encounters with his own race he remains a stranger to this most human capacity. As out of character as the laughter itself, there is a note of hubris in it.[38]

Human and Other Relationships

Tarzan is extraordinarily loyal, often to individuals unrelated to him or undeserving, but rapport with his fellow beast comes more naturally to Tarzan than his often-reluctant tendency to become involved in human affairs. His conscience in a sense drags the resisting ape-man into the society of humans, but animals are a different matter. "Numa [the lion] and Sheeta [the leopard] he admired — his world would have been desolate without them; but the men who were his enemies he held only in contempt. He did not hate them. Hate was for them to feel in their small, warped brains. It was not for the Lord of the Jungle."[39]

Tarzan is at his most interesting when he is reacting to an animal as one of them. In this aspect Tarzan appears to be unique in his lineage. He always respects animals, whatever they are to him. Ever ambivalent, never arrogant or smug in his humanity, Tarzan is forever unpredictable.

His disaffection for his fellow humans can be poignant, as when he expresses it to d'Arnot (in a sense for us, as well as on his own behalf): "Paris is no place for me. I will but continue to stumble into more and more serious pitfalls. The manmade restrictions are irksome. I feel always that I am a prisoner. I cannot endure it, my friend, and so I think that I shall go back to my own jungle, and lead the life that God intended that I should lead when He put me there."[40]

In short, Tarzan's unique charisma depends upon his duality. The ape-man says it best when Jane marvels at his ability to train a lion cub (and he is distancing himself from his own wife or mate in the process): "Your kind are not afraid of you — these are really my kind, try to civilize me as you will, and perhaps that is why they are not afraid of me when I give them signs of friendship."[41]

On more than one occasion Tarzan expresses his disgust with self-styled civilized humankind. In at least one situation his response to humanity is a placid indifference mingled with no more than mild curiosity. Understandably, Tarzan's friends are few and, like d'Arnot, selective.[42] Tarzan's disaffection for collective humanity remains with him for a lifetime. At one point he dissuades another from returning to civilization, noting, "I . . . have left my jungle and gone to the cities built by man, but always I have been disgusted and been glad to return to my jungle — to the noble beasts that are honest in their loves and in their hates — to the freedom and genuineness of nature."[43] Thus, a little civilization goes a very long way, and the jungle is Tarzan's true home.

As compelling as are Tarzan's characteristic reactions to his fellows, their responses to him in many ways reveal more of the man. It is telling that the interplay with the beasts, while vividly portrayed, is far less complex than that with humanity. To his fellow beasts Tarzan is of the jungle. He is *mangani* (ape) to them, never *tarmangani* (Caucasian human). Unfortunately for the ape-man, he "grows" away from his people, the mangani, even before his first contact with Europeans and Americans. When later returned to the jungle, he turns to the animals, not people, for companionship.[44]

Except for his enemies who all fear him, people inevitably recognize him as a leader. Even when unconscious, Tarzan is impressive. A friend sums him up by comparing him to a lion, with his strength, majesty, and ferocity, "[i]t was a controlled ferocity, but it was there — Wood felt it. And that, perhaps, was why he was a little afraid of him."[45]

Women, from Jane to Nemone and especially La of Opar, inevitably fall in love with Tarzan.[46] What sets Burroughs and Tarzan apart from their genre in this regard is the relative subtlety of the character as reflected in the women who yearn for him. Bertha Kircher, a sophisticated woman of the world, speaks for them all from the civilized perspective. In a way she is revealing the compelling nature of Tarzan in all civilized eyes: "She could not fathom him. Never in her life had she seen a man at once so paradoxical and dependable. In many of his ways he was more savage than the beasts with which he associated and yet, on the other hand, he was as chivalrous as a knight of old."[47]

Ambivalence

If there is a single word that describes Tarzan as Burroughs presents him, that word is "ambivalent," in association with freedom-craving, steadfast, fearless, and, occasionally, lonely — all serving to render him a timeless surrogate for us as we maintain our own balance and interactions with human and beast.

Domestication, unfortunately, can be a detriment to any creature of the wild, and Tarzan is no exception. Human emotions and motivations are perhaps this man's worst enemies. While still in the jungle and no more than vaguely aware of his human side, Tarzan often yearns for something that he knows is lacking in his existence. There is something for which there is no outlet. It is defined by Burroughs as sentient companionship — a boon only civilization can offer. And, yet, even when the association with humans has been prolonged, Tarzan seldom considers himself a man, a creature he includes among his blood enemies. In this response especially, the ape-man is profoundly ambivalent and remains so for a lifetime. "Strong within him surged the jungle lust that he had thought dead."[48]

Except for his steadfast love for Jane, an absolute need for freedom is the driving force behind Tarzan's every move. But equally compelling to him and to his audiences, Tarzan's dual nature renders him not merely alone; he is frequently lonely. His first inclination upon a return to the jungle is to take to the trees.[49]

Loneliness is an inevitable companion of Tarzan's ambivalence with regard to mangani and tarmangani ways. His friendships and, especially, his love for Jane can transcend both the loneliness and, to some extent, that ambivalence. These relationships alone can blend the ape-man's two persona into a whole greater than its parts. "For a moment Tarzan stood irresolute, swayed by conflicting emotions of loyalty to D'Arnot and a mighty lust for the freedom of his own jungle. At last the vision of a beautiful face, and the memory of warm lips crushed to his dissolved the fascinating picture he had been drawing of his old life."[50]

Over time the novels reveal three powerful figures in one: Tarzan is a leader among his apes, he is Lord Greystoke among Europeans and the occasional American, and chief of the Waziri people. In large part those qualities of power and ability result from his unique history, and Tarzan learns to utilize that facet of his multiple natures most appropriate to the circumstances in which he finds himself. But, however civilized Tarzan becomes, he is always on the brink of reversion to mangani even though he can no longer know contentment among the beasts. His mangani self demands frequent reinforcement.[51]

There is no denying at least a superficial appeal in Tarzan's wild freedom: "The ape-man's mind was untroubled by regret for the past, or aspiration for the future. He could lie at full length along a swaying branch, stretching his giant limbs, and luxuriating in the blessed peace of utter thoughtlessness, without an apprehension or a worry to sap his nervous energy and rob him of his peace of mind. Recalling only dimly any other existence, the ape-man was happy. Lord Greystoke had ceased to exist."[52]

The dark side of Tarzan's ambivalence is his frequent contempt for things civilized. He is never surprised by the perfidy of humankind.[53] His human enemies can drive the ape-man to near-madness: "Tarzan felt the old, wild impulse to reply with the challenge of his kind. His kind? He had almost forgotten that he was a man and not an ape. . . . He felt a wild wave of madness sweep over him as his efforts to regain his liberty met with failure."[54] "Tarzan is not as other men; the training and instincts of the wild beast have given him standards of behavior and a code of ethics peculiarly his own."[55]

MANIFESTATIONS

The phenomenal and continuing appeal of Tarzan in his original form tells us much about contemporaneous needs throughout this century and all over the world for a wild man who is also a noble savage with clear hints of the Hero and the Wild Man. But it does not stop there. Tarzan is compelling to audiences who have never heard of the books. That popularity is attributable to the cinematic images of Tarzan. Even though substantially reduced in complexity by these manifestations, the ape-man still provides something the world needs.

In the Cinema

Almost all the cinematic Tarzans, however disappointing to the fans of Burroughs's complex character, have proven phenomenally successful in their own right. Four cinematic eras can be discerned: the early version relatively closely supervised by Burroughs; that of Johnny Weissmuller; the interim versions purportedly attuned to the times from the 1960s into the early 1980s; and the naturalist version of Hugh Hudson, calling to mind *The Wild Child* of Itard and Francois Truffaut.[56] At this writing a fifth era is imminent, with no less than a pilot and new television series in the fall of 1996.

For the most part film has failed to portray the ape-man in all his complexity. One line emphasizes the wildness to the exclusion of the man's civilized persona, while another focuses on the civilized to the exclusion of the wild. Still, something of that duality and ambivalence must be conveyed and strikes a chord in many a person. Burroughs was and most of his fans continue to be distressed by the fact that Weissmuller *is* Tarzan for the vast majority of the population. This Tarzan suffers from a childish, rather than childlike nature and, unlike the original, is completely baffled by any sophisticated interaction.

Two versions of Tarzan following Weissmuller came closer to the original but still presented only a portion of the complex persona. They are Lex Barker and Gordon Scott. The former was possessed of an aristocratic bearing, brooding silence, an animal magnetism, and little material with which to work. Gordon Scott, while excessively muscular for Burroughs's version, at least in one film portrayed an articulate but taciturn Tarzan with his own unique responses to the death of a friend, the need to bring the killers to justice, and the way to deal with a spoiled socialite.[57]

Following Gordon Scott's version, producers attempted to render the ape-man "relevant." The result was mixed, with Jock Mahoney essaying a civilized man who happened to live in the jungle while engaging in good deeds around the world. He was a little too civilized, and the stories lacked the element of a man who returns to his own jungle to refresh his spirit. Worse yet was Mike Henry who lacked rapport with the animals and who was too wooden an actor to convey any sense of mystery about the man. Moreover, in the era of Sean Connery's James Bond, there was an effort to render Tarzan a kind of wilder version of 007. The John and Bo Derek remake[58] of the original MGM Johnny Weissmuller version marks the end of that era and the beginning of one in which this literary character is claiming serious attention as a classic image. The film had its audience but was a failure on numerous levels. The advertising campaign is the highlight of this effort: "The most beautiful woman of our time in the most exotic adventure of all time," often displayed against a still of Jane in Tarzan's arms as he swings them both from a vine. Miles O'Keeffe is as stalwart as any Tarzan should be and Bo Derek's expression of delighted wonder with just a touch of fear is a picture right out of the original novel.

The most interesting recent version of Tarzan is that of Hugh Hudson, *Greystoke: The Legend of Tarzan, Lord of the Apes* (1984). Early commentary suggested that Burroughs's creation was finally to be realized on film. Hudson was reportedly entirely aware of Tarzan's significance to a large audience in that Burroughs had "chanced on an elemental and almost endlessly resonant myth in his invention of Tarzan."[59]

The end result was a serious retelling of Burroughs's version of the ape-man's origins with a subtle sense of humor, but the adventure was almost exclusively of the mind. Once Tarzan reached maturity and succeeded to lordship of the apes, he encountered no more physical conflicts. In fact, he backed off the few that were threatened by others. In short, in this film too much of the best of Tarzan's character and unique strengths yielded to the internal ambivalence. Only modest hints of the ape-man's courage, jungle craft, and ability to adapt at least superficially to the ways of civilization were revealed. This portrayal, while closer to the original than most, still missed the mark as had its cinematic predecessors. Nevertheless, it was well received by critics and audiences alike.

On Television

Announcements of the live-action television series in the mid-1960s were widely welcomed, but Burroughs fans suspected a serious catch threatening the very nature of Tarzan that makes him so relevant over so long an interval: First of all, there was Tarzan's "ward," Jai, to be protected from all manner of evil. Then there was the deliberate ambiguity about location. Jane was among the missing, as was Tarzan's wild nature.[60]

Even *The Wall Street Journal* took note of the coming event.[61] This article predicted a surge of new "Tarzania" and quantified the extraordinary popularity of the character, reporting that television was running the thirty-six movies five thousand times per year and noting that in New York City Tarzan reruns garnered a 35 percent larger audience than the average late-night film. Actually, the television animated version was more faithful to the Burroughs canon.[62] In short, a flippant, Americanized, "educated" Tarzan, though popular enough to survive three seasons, cannot replace Burroughs's complex English peer and lord of the apes.

RECEPTION AND LONGEVITY

For all his international popularity and phenomenal longevity, Burroughs's Tarzan has not been well received by contemporaneous critics. Many of them, it seems, neglected to read what they were criticizing. One even denigrated Tarzan's origin story without reading it and proceeded to praise the Johnny Weissmuller version.[63]

In the 1960s

The mid-1960s was a time of amazing resurgence of interest in Tarzan, and evaluations of the novels took on a more balanced tenor. One of the initial articles referred to the ape-man as the most famous fictional hero of this century, possibly including film, television, Sunday comics, and comic books in that assessment. Included in the praise was the narrative power of a great storyteller able to catch readers up and weave a web about them holding them to the last page. The public loved the action filled, blood and thunder writing.[64]

Fifty years after Burroughs had first published, excerpts were beginning to appear in American and European textbooks. Tarzan himself was tentatively acknowledged as "the only great legendary hero produced by the literature of this century."[65]

Analyses of the Books

It was not long before a series of entirely serious books commenced analyzing Burroughs and, especially, Tarzan.[66] Richard Lupoff's book is an unabashed tribute to Burroughs's writings in which the endcover acknowledges Tarzan as his most famous creation and a household name around the world. Tarzan's popularity is attributed to his representation of "manliness, strength, courage, and perhaps just a touch of bestiality."[67] Whatever his literary origins, Tarzan is described as fascinating:

No simple primitive here, but a complex figure of introspective nature, a man whose development — in contrast with the jungle, then with the artifacts of civilization, and finally with the representatives of humanity — is painstakingly detailed. By virtue of Tarzan's natural intelligence he could never feel fully comfortable living the primitive bestial existence of his boyhood and youth. And yet those jungle years . . . had taught [Tarzan] a different code, a more honest and simpler code than that of man. A code and an outlook that would prevent him from fully acclimating himself to human society.

Partaking of two worlds, never fully at home in either, the melancholy figure of Tarzan stands far above the simple jungle adventures of screen and cartoon page.[68]

Lupoff takes note of the large body of descendants and pastiches of Tarzan, both legitimate and pirated,[69] yet another indication of the extraordinary popularity of feral-man images.

During the 1980s the term "classic" began to creep into discussions of Burroughs's Tarzan, with comparisons to the Golden Age of Greek literature. Erling Holtsmark notes that

Tarzan himself is a surprisingly complex literary persona whose clear roots in the mythological heroes of antiquity, notably, Odysseus, are combined with features borrowed from American Indian traditions. An examination of Tarzan's relationship with animals serves to place him in the larger context of heroic literature and its divine machinery.[70]

[He is] the most famous fictional character who links man to his evolutionary past. [Like the Greek and Roman heroes who manifest] the transitional relationship between gods and men . . . by personal example . . . , Tarzan stirs archetypal recollections of what we today believe to be our own origins. . . . There are many reasons for Tarzan's great popularity, but possibly this aspect of the character as the "missing link" is one of them.[71]

Holtsmark takes Tarzan seriously indeed and sums up his relationship to literary heroes in the observation that, while Burroughs generally maintains a sharp distinction between the worlds of human and animal, it is Tarzan alone "who, like the semi divine heroes of classical literature and myth, can straddle the two worlds."[72] And yet Tarzan is hardly bound by any such conventions. His utter freedom and capacity to be whatever he chooses strikes a chord in readers everywhere and across nearly a century. According to Holtsmark, "He combines the best of the man who is physically active and intellectually inquisitive, and both possibilities are part of his British aristocratic roots. Tarzan is able to bridge the two worlds of knowledge and action, for in understanding the one in terms of the other, he has unified these two primary areas of human experience, word and deed, thought and action, mind and body."[73]

Add to those dichotomies civilization and wildness, and you have the summation of the character's lasting appeal. And yet he has retained his humanity, as most of us would choose to do, and this element of reality renders him more approachable for those who would identify with him. Tarzan agonizes but he triumphs, carrying us with him. For all his freedom, Tarzan — in keeping with the classical tradition — suffers from the very traits that render him heroic.[74]

Cinematic Transition

Despite the questionable efforts of the 1960s and 1970s to convert Tarzan into a James Bond clone, critics were at least attentive. An article in *Life* reported *Tarzan's Three Challenges* and *Tarzan Goes to India* with acknowledgment not to Burroughs but to the Weissmuller legacy. The author asked whether this courtly, soft-spoken, shrewd, excessively refined man could be Tarzan. Commercial success notwithstanding, the author was wistful, "admirers of the old Tarzan may feel that the changes wrought on his character have . . . injured the ape man beyond repair."[75]

It was in the era of the heroic rediscoveries of George Lucas and Steven Spielberg that both "camp" and James Bond were left behind and new versions of Tarzan were presented. The first was the roundly criticized Bo and John Derek remake of MGM's *Tarzan the Ape Man*. Perceiving the missed possibilities, neither Burroughs fandom nor mainstream critics accepted this effort.[76]

Both *Newsweek* and *The Wall Street Journal* reviewed the film, an indication of the popularity of the character despite his lengthy absence from the screen. The former reviewer implies a longing for something closer to the Burroughs version of the character, with the comment that this Tarzan never speaks, smiles, or indicates any suggestion of personality. In short, he's some hunk, and that is all the Dereks have to say about the Lord of the Jungle.[77] *The Wall Street Journal* summed it up in the terms "insufferable" and "ineptitude."[78] Why these publications would even bother with the newest Tarzan film is expressed in Gene Siskel's review: "Actually, it's a shame that . . . John Derek didn't make 'Tarzan' a more blatantly erotic movie. The Tarzan story is a *classic fairy tale* that could have honestly delivered some very heavy breathing" (italics added).[79]

The prospect of Hugh Hudson's effort was much better received.[80] Hudson perceived Tarzan not as a musclebound oaf, but as a man who was light and lithe, every muscle balanced through use in his active life. Echoing Burroughs himself, Hudson clearly perceived the classicism represented by the character and promised to bring out not one but two journeys of self-discovery; Tarzan's and, surprisingly, Paul d'Arnot's rather than Jane's. For all its many flaws, *Greystoke* succeeded on more levels than any of its predecessors. In fact, early scripts suggest an effort far closer to Burroughs's images than was finally realized.[81] One has to wonder how much more popular the film might have been, had John Clayton been shown to be as effective in society as he proved to be among the apes. The filmed version cut moments revealing Tarzan's phenomenal prowess in the jungle and rendered a distinctly uncomfortable Lord Greystoke. Only at the conclusion as filmed does Tarzan, upon returning to the jungle, pause for a moment, clearly torn between his two worlds.

The numerous reviews of *Greystoke* were respectful if not entirely positive. Prior to its release an article in the *Chicago Sun-Times* distinguished this new effort from all its predecessors, notably the Derek effort.[82] Unfortunately, the article perpetuated the mistakes resulting from the comparison to the earlier films instead of the original books. The review in *Time* was more accurate.[83] Roger Ebert's review also acknowledged the numerous bad versions he identified as "mythological schlock" before observing, "Yet the *idea* of Tarzan remains indestructable."[84]

On Primetime Television

The 1960s television version of Tarzan drew the critics and a series of contemporaneous entertainment icons as guest stars, not the least of whom was Ethel Merman. One of the associated reviews indicated some knowledge of the original books and refers to Tarzan or Lord Greystoke as "that greatest of all civilization-escapers."[85] While it is informative that the media were interested in this version of Tarzan, the serious students of the ape-man wanted something closer to what makes the ape-man so significant. After a while the stories became more adult and Ron Ely adapted some characteristics from the books.[86] For all its flaws, however, the series actually did better in the ratings than the contemporaneous *Star Trek.*[87]

NOTES

1. The twenty-plus individual novels (including the posthumous publications of the 1960s) have been published in a variety of editions. Those cited here are all Grosset & Dunlap editions unless otherwise noted: *Tarzan of the Apes* (1914, "*Tarzan*"); *The Return of Tarzan* (1915, "*Return*"); "*The Beasts of Tarzan* (1916, "*Beasts*"); *The Son of Tarzan* (1918, New York: Burt; "*Son*"); *Tarzan and the Jewels of Opar* (1918, "*Opar*"); *Jungle Tales of Tarzan* (1919, 1963 edition, New York: Ballantine Books; "*Tales*"); *Tarzan the Untamed* (1920, "*Untamed*"); *Tarzan the Terrible* (1921, "*Terrible*"); *Tarzan and the Golden Lion* (1923, "*Golden Lion*"); *Tarzan and the Ant-Men* (1924, "*Ant-Men*"); *Tarzan, Lord of the*

Jungle (1928, *"Lord"*); *Tarzan and the Lost Empire* (1928–1929, New York: Metropolitan; *"Lost Empire"*); *Tarzan at the Earth's Core* (1929–1930, *"Earth's Core"*); *Tarzan the Invincible* (1930–1931, *"Invincible"*); *Tarzan and the Lion Man* (1934, *"Lion Man"*); *Tarzan and the Leopard Men* (1935, ERB, Inc.; *"Leopard Men"*); *Tarzan's Quest* (1936, ERB, Inc.; *"Quest"*); *Tarzan and the City of Gold* (1938, Tarzana, CA: ERB, Inc.; *"Gold"*); *Tarzan and the Forbidden City* (1938, ERB, Inc.; *"Forbidden City"*); *Tarzan and the "Foreign Legion"* (1947, ERB, Inc.; *"Foreign Legion"*); *Tarzan and the Castaways* (1964, New York: Ballantine; *"Castaways");* *Tarzan the Magnificent* (1964, Ballantine; *"Magnificent"*); *Tarzan and the Madman* (1964, Ballantine; *"Madman"*).

For more on Tarzan as Hero, see Slate, "Edgar Rice Burroughs and the Heroic Epic" (1968), 118.

2. *Tarzan.*
3. *Return.*
4. See, e.g., *Lord.*
5. See *Tarzan*; *Return*; *Lord.*
6. See *Beasts*; *Untamed*; *Golden Lion*; *Invincible.*
7. *Terrible*, 118.
8. *Tarzan.*
9. *Gold*, 15.
10. *Golden*, 65; *Lion-Man*, 299.
11. *Tarzan*, 209.
12. *Magnificent*, 7.
13. *Jungle Tales*, 159–160 .
14. *Terrible*, 89.
15. *Untamed*; *Gold*, 37.
16. *Lion-Man*, 241.
17. See, e.g., *Terrible*, 119; *Core.*
18. *Return.*
19. See, e.g., *Ant-Men*, 76.
20. *Opar*, 146.
21. *Jungle Tales*, 113.
22. *Untamed*, 9–10; *Beasts*, 13.
23. *Tarzan*, 178.
24. *Lion-Man*, 102.
25. *Gold*; *Lion Man*; *Jungle Tales*, 135.
26. *Return.*
27. *Beasts*, 32, 34–35.
28. *Gold*, 289; *Golden Lion.*
29. *Return*, 36, 37–38, *Untamed, passim.*
30. *Untamed*, 20.
31. *Gold*, 82.
32. *Tarzan*, 281.
33. *Jungle Tales*, 42.

34. *Magnificent*, 22–23.

35. *Gold*; *Ant-Men*.

36. *Lost Empire*, 199.

37. *Lord*; *Invincible*; *Leopard Men*; *Quest*; *Foreign Legion*; *Ant-Men*.

38. *Jungle Tales; Quest; Lord; Castaways.*

39. *Madman*, 8.

40. *Return*, 73.

41. *Golden Lion*, 8.

42. *Lion Man*; *Magnificent.*

43. *Golden Lion*, 242–243.

44. *Lost Empire*; *Tarzan*; *Beasts.*

45. *Lord*; *Earth's Core*; *Golden Lion*; *Untamed*; *Magnificent*, 27.

46. *Tarzan* (Jane); *Return* (La, the Oparian high priestess); *Untamed* (Bertha Kircher aka Patricia Canby, a British agent); *Gold* (Nemone, the cruel and insane queen); *Lion-Man* (Rhonda, the film star); *Lord* (Guinalda, a medieval noble woman); *Invincible* (Zora, a putative communist agent); *Forbidden City* (Magra).

47. *Untamed*, 191–192.

48. *Beasts*; *Jungle Tales*; *Forbidden City*; *Magnificent*; *Son*, 29.

49. *Tarzan*, 281.

50. Ibid., 282.

51. *Foreign Legion*; *Return*; *Untamed*; *Son*; *Earth's Core.*

52. *Opar*, 32.

53. *Untamed*; *Magnificent.*

54. *Return*, 132–133.

55. *Lion-Man*, 129.

56. See Chapter 7 for Victor as described by Itard himself. The film is *L'Enfant Sauvage* (1969, Francois Truffaut). The English version is available on videotape.

57. *Tarzan's Greatest Adventure* (1959, John Guillermin). For more on the Tarzan films, including television, see Gabe Essoe. *Tarzan of the Movies: A Pictorial History of More than Fifty Years of Edgar Rice Burroughs' Legendary Hero*, 1968, New York: The Citadel Press, and David Fury, *Kings of the Jungle: An Illustrated Reference to "Tarzan" on Screen and Television*, 1994, Jefferson, N.C.: McFarland & Company, Inc.

58. *Tarzan, the Ape Man* (1981, John Derek).

59. Nightingale, "After 'Chariots of Fire,' He Explores the Legend of 'Tarzan,'" *New York Times*, March 6, 1983, II17.

60. Val Adams, "Tarzan Coming to TV without Jane," reprinted in *The Gridley Wave* 20 (May 1966), unpaginated.

61. Ronald Buel, "The Swinger's Return: Tarzan (but Not Jane) in a Jungle Comeback," *The Wall Street Journal* 1966.

62. Maloney, "Edgar of the Apes: He Had the Craziest Dream — and 'Tarzan' Was Born," *TV Guide*, October 22, 1966.

63. Birrell, "The Glories of Excess," *The New Statesman and Nation*, May 21, 1932, 661.

64. McGreal, "The Burroughs No One Knows" (1965), 12, 14.

65. Ibid., 15.

66. Lupoff, *Edgar Rice Burroughs: Master of Adventure* (1965); Holtsmark, *Tarzan and Tradition: Classical Myth in Popular Literature* (1981); and Lupoff, *Edgar Rice Burroughs* (1986).

67. Lupoff, *Edgar Rice Burroughs*, 3.

68. Ibid., 170, 209.

69. Ibid., 236.

70. Holtsmark, *Tarzan and Tradition*.

71. Ibid., 83.

72. Ibid., 70–71.

73. Ibid., 109.

74. Ibid., 112. See also Torgovnick, *Gone Primitive* (1990) for a further expansion of the cultural relevance of Tarzan.

75. Orshefsky, "Gone Are the Pals of Yesteryear," *Life*, June 14, 1963, 102.

76. Corriell, "Tarzan the Ape Man/Me Bo — Me Show," *The Gridley Wave* 83 (Summer 1981), unpaginated.

77. "Lost in the Jungle," *Newsweek*, August 31, 1981.

78. Boyum, "New Films: Stevie and John, Arthur and Tarzan," *The Wall Street Journal*, 1981.

79. Siskel, "'Tarzan': Innocent Bits of Hollywood and Vine," *Chicago Tribune*, July 27, 1981, Sec. 2, 5.

80. Nightingale, "After 'Chariots of Fire,' He Explores the Legend of Tarzan."

81. Drafts attributed to Robert Towne, one dated August 4, 1977, and two others identified as drafts 4 and 6 and dated 1982.

82. Caulfield, "He 'Greystoke,' They Not 'Tarzan,'" March 3, 1983.

83. Schickel, "The Wild Child Noble Savage," *Time* (1984), 89.

84. *Chicago Sun-Times*, March 25, 1984.

85. "Merman of the Jungle," *TV Guide*, September 16, 20.

86. "Short Waves," *The Gridley Wave* 22 (1967), 2.

87. *The Gridley Wave* 25 (1968), 3.

14

Related Contemporaries and Descendants of Tarzan

RELATED CONTEMPORARIES

Mowgli and Tarzan are near contemporaries, but Tarzan's heroics — and his ambivalence — begin where Mowgli's story closes. Unlike Mowgli, who is left as the *man-cub,* the child of humankind, Tarzan becomes and remains for a lifetime an *ape-man*, forever balancing the two identities, wild and civilized. Outstanding among popular noble savages contemporaneous with Tarzan are Conan and Kioga. They are less compelling than Tarzan, because Conan is mostly savage and Kioga mostly civilized. Neither Conan nor Kioga is feral; with both the precarious balancing of two identities is sorely missed. Conan is a young barbarian of a particularly savage nature quite acceptable in his time. Kioga, nurtured within a lost tribe of Amerindians, is a Caucasian who turns to the forest for companionship when, as a child, he is not accepted among his adopted peers.

Conan the Barbarian

Conan the Barbarian was created by Robert E. Howard during the first half of the twentieth century.[1] Conan is of the heroic fantasy, sword, and sorcery genres, which underwent a revival in the 1980s. While the Burroughs influence may well have been there, these tales, along with Burroughs's, hark back to primitive peoples and ancient times — atavistic myths, legends, and epics. Conan is a superhero and a modern version of the classical Hero.[2] There is a wildness about him: "The [horses] were close to nature. So was Conan. . . . Like his, their senses were more delicately turned to the aura of evil than were the senses of city-bred men." "Conan woke suddenly. Some eerie premonition — some warning from the barbarian's hyperactive senses — sent its current quivering along the tendrils of his nerves. Like some wary jungle cat, Conan came instantly from deep, dreamless slumber into full wakefulness. He lay without movement, every sense searching the air around him."[3]

Conan is close to nature in the sense of the wild beast; it is both his nature and his nurture. Because he is not feral he does not long for something he does and simultaneously does not possess. His straightforward character lacks ambivalence. While popular, he has not been the subject of the profound, worldwide fascination evoked by Tarzan, perhaps because Conan is too far removed from us and our own ambivalence and longings for some element of wildness while retaining our civilized natures.

Conan suffers fits of melancholy gloom, the origin of which is not apparent. While he fears neither man nor beast, he does indulge an atavistic terror of the supernatural that can actually leave him frozen in horror and even frantic: "At heart a superstitious barbarian, he feared nothing mortal but was filled with dread and loathing by the uncanny supernatural beings and forces that lurked in the dark corners of his primeval world."[4]

In the resurgence of interest in Conan through the writings of L. Sprague De Camp and Lin Carter, Conan becomes almost as feral as primitive. He is described as naturally taciturn, neither asking for nor accepting companionship. He moves "with the feline caution of a stalking leopard."[5] When abused by enemies Conan becomes even more dangerous in a retreat into something bestial: "His tangled hair fell ropelike around his battered face, a cracked and sunburnt mask in which only the eyes lived. They were the angry, burning eyes of a trapped and dying beast."[6] Such references to a beastlike nature are constant.

Like Tarzan, this Conan possesses a profound need for freedom and would prefer the honorable death of battle to captivity. "Unarmed, he harbored no illusions about the outcome; but better to sell his life dearly than to present a willing neck to axe or knotted rope." And his pride is natural and innate, without regard to any image as is too often the case with pastiches of Tarzan.[7] At least in this portrayal Conan remains relentlessly primitive without feral ambivalence. He indulges in excess in more than one endeavor and possesses no tempering civilization. He is neither overly noble nor noted for his chivalry.

Kioga of the Wilderness

Kioga, every inch the *noble* savage, is the antithesis of the shaggy, barbaric Conan. The four tales of Kioga were written in the mid-1930s.[8] Kioga is superficially a fascinating study, but he seems perfectly at home wherever he finds himself, his romance with the heroine is left undeveloped, and he spends even less time in civilization than does Lord Greystoke. Several passages, all from *Kioga of the Unknown Land,* suggest this character's strengths for present purposes:

[B]orn to rove a wild hawk of the wilderness, indeed an outcast with only shaggy forest denizens for company, until qualified by his fighting sinews to become the greatest warrior in all the Shoni tribes.

[His] was a figure to command interest, envy and admiration: straight, tall, formed like a living bronze. His splendid torso was naked to the waist; in one hand was a hunting bow, and at his back a quiver full of arrows. Here was the ideal form of the hunter-warrior, combined with the features and expression of a well-browned civilized white man.

Fine, clean-cut features on a dark and shapely head, a pair of green-blue eyes with fiery intrepid gaze, a torso bronzed as any Indian's. Wide-shouldered, lean of waist, long-limbed — with coil-like muscles indicating almost boundless strength and bottomless endurance.[9]

Among his lapses for present purposes, while alone much of the time, Kioga is more solitary than lonely. While he and his wilderness are undeniably evocative, the duality and associated ambivalence are lacking: "A forest stretched before us toward a misted valley. In the crowns of several trees hawks had their nests; far in the blue great condors circled slowly. This was Kioga's favorite solitary realm, up where only hawks and eagles dwelt."[10]

Kioga is, however, a complicated mix of loyalties, given his American parentage, his nurturing among the Shoni, and his adoption by bears. Although made unhappy among the Shoni, Kioga returns to them out of loneliness and makes a respected place for himself among them, only to find tribal life too confining.[11] Unlike Tarzan, Kioga is actually outlawed by the Shoni when framed for the murder of his foster parents. Hence, he is an outcast.

Like Tarzan, Kioga bears a streak of cruelty and can be vindictive. Unlike the ape-man, however, Kioga takes a certain joy in this cruelty. He loves to kill and will use human tools to punish an animal. As a result, he is hated by his animal enemies, but, again like Tarzan, Kiogo counts man as his "direst" enemy.[12]

When among his civilized people Kioga is much like Tarzan in his response to the restraint frequently imposed by society, legitimate or otherwise. During a holdup in a New York restaurant, the criminals single out Kioga (Lincoln Rand): "Ah, but here was no tame and submissive prey, awaiting an outlaw attack in fear and trembling! Lincoln Rand, thus far, had submitted to the training collar of civilization; but now the collar slipped. As that gun came up, Lincoln Rand sloughed off name and white-man's teaching and became in an instant the Snow Hawk, the primitive cliff-man disturbed at his meat. He struck as he had learned to strike — fast, hard, suddenly."[13]

When Kioga resists the police in "the unreasoning reaction of the wild . . . animal whose sole obsesssion is to evade a trap," he is arrested. During his incarceration, prolonged by his refusal to communicate, Kioga reaches the realization that the whole of civilization is a prison, and he disappears when he is finally freed.[14] Thus, he returns to the wild where his nobility reigns without the aura of "otherness" an attempt to bridge his two worlds would instill.

Terangi of the South Pacific

Terangi is the reverse of Tarzan and Kioga. He is by birth the noble savage, who attempts to become a part of the civilized world with tragic result. His tragedy is, however, ultimately overcome through Terangi's own inner strength and with the kindly assistance of a few noble souls among the civilized players. Terangi's people are of an isolated South Pacific island: "The inhabitants, few in number, are Polynesians, with the cheerful dignity of their race, but the loneliness, the enforced simplicity, and the precariousness of life faced with the perpetual menace of the sea have made them sturdy and resourceful, and have implanted in them an abiding sense of the tragic nature of man's fate."[15] Among them Terangi carries a "sense of dignity not to be affronted without risk."[16]

Unfortunately, Terangi's dignity *is* affronted, and he quite naturally retaliates upon provocation. Because he is not of the accepted race, he is jailed. His multiple escape attempts are all thwarted, and the brief term expands to an appalling sixteen years. It is his unquenchable craving for freedom in his own wilderness that is deeply evocative.

He showed an ingenuity and a fierceness of determination in getting away that were new to the experience of the police. . . . As soon as he was given a measure of freedom within the walls, he would find a means of getting outside them.

Although he felt keenly the injustice of his first imprisonment, he was too much of a man to hoard up bitterness. He knew that his captors were doing no more than their duty and nursed no resentment toward them. But he had to be free, whatever the cost.[17]

In Terangi's final escape, a man is killed. Somehow Terangi manages to sail alone back to his home island. In preparation for the coming hurricane he encounters the wife of the local official whose unbending sense of duty has precluded Terangi's being freed: "Madame de Laage stopped short at the sight of him, then her glance met mine. . . . Both astonishment and relief were in her glance, but when she again looked toward Terangi there was no light of recognition in her eyes. She was determined not to see him."[18]

When the hurricane ultimately passes, there are very few survivors. Throughout the ordeal Terangi's nobility is demonstrated repeatedly. Madame de Laage is one who survives only because of his protection. Even now Terangi and his wife, both strong swimmers, remain with the woman until her rescue is imminent. Upon being reunited with his wife, de Laage spots the distant canoe bearing the fugitive to safety. With an impulsive inspiration of the humane over his own narrow perception of duty, de Laage observes that it is only debris. Justice prevails at last, and Terangi will be allowed his precious freedom after all. In the years to come he is granted a pardon.[19]

DESCENDANTS

The fact that, after nearly a century, Burroughs's Tarzan continues to be manifested and represented in a variety of formats is a testament to the hold he maintains on the imagination. The quality of the work bearing his name may be immaterial. His name alone draws interest. Tarzan also has inspired numerous intellectual descendants. Some are legitimate, others not, but all are a further indication of our need to find surrogates in the expression of our ambivalence in choosing wildness or civilization.

Lineal Descendants and Other Close Relatives

Parodies

The perennial popularity enjoyed by Tarzan is reflected in the comic touch, subject to the control exercised by the intellectual property proprietor, Edgar Rice Burroughs, Inc. Usually a pseudonym is used, such as

Jimmy Durante's "Schnarzan," an early interpretation.[20] *Mad Magazine* established Melvin of the Apes, known to *Mad*'s readers for his chilling cry of "Hoo Haa!" As might be expected, the parody is more of Weissmuller's interpretation than of Burroughs's. Like Tarzan, Melvin enjoys an astonishing rapport with his fellow creatures. When his "hoo haa" of distress rings out, the response brings on the expected elephants and apes — along with whales, pigs, and even a few extinct types. There is a dinner scene that comes to mind upon viewing *Greystoke*'s counterpart, although the *Mad* version is played for the broader laugh.

Television has offered *George of the Jungle*. With its reference to one "Kerchak," this parody suggests a higher regard for the Burroughs originals than many other versions of the ape-man.[21] Unforgettable is the image of Jack Benny as Tarzan and Carol Burnett as Jane, growing old together somewhat less than gracefully. The most sustained spoof is *The World's Greatest Athlete*, a book and film of Walt Disney Productions.[22] Lest anyone miss the connection, the heroine is "Jane" and predictably, she has a profound effect on "Nanu's" assimilation into society. For all the physical humor, the visual images are extremely close to those projected in the Weissmuller and other films.

Legitimate Imposters

There are very few legitimate literary "imposters," again their very existence is suggestive of the enduring popularity of the character. One of the most interesting is a "novelization" of the Weissmuller *Tarzan the Ape Man*, a compilation of Rex Maxon's daily comic strip issued by the House of Greystoke (imprint of the Burroughs Bibliophiles). Not surprisingly, the story line includes a substantial effort to reconcile the cinematic Tarzan with the literary. Arthur B. Reeve's novelization of *Tarzan the Mighty* strives for a similar reconciliation.[23] Even with subtle references to Tarzan's keen senses and intelligence, this material is pallid in comparison with the originals. *The New Adventures of Tarzan* fares not much better even though the ape-man's civilized persona is allowed to emerge.[24] This is perhaps the only instance outside the novels in which Tarzan's civilized half is portrayed without loss of the wildness or the dignity inherent — and crucial in terms of his longevity — to the ape-man.

An interesting but disappointing novelization comes out of the cinematic era of James Bond's early popularity.[25] Here again was an attempt, almost certainly unnecessary, to render Tarzan palatable to the audience of the time. Here he is reduced to one more sophisticated adventurer, albeit with a unique talent for getting about in the jungle. The effort, apparently a respectful one, suggests that pastiches of this character simply do not work, especially when there is little or no effort to work in Burroughs's truly unique style.

A major failure in such efforts is in the inability to portray effectively one side of the character's persona or the other. In this case Tarzan is overly civilized with no more than a veneer of wildness. Nonetheless, there is the tantalizing reference to the all-important symbiosis between those two persona: Tarzan is "unique among men, on the thin borderline between the animal and the human, gifted with the natural instincts of the former and the wisdom of the latter." There is an interesting split between the two in the era's compulsory car chase, and there is an intriguing aura about this Tarzan, "a sense of depths and powers, a savage electricity, that excited, disturbed, and even frightened."[26]

Upon making the transformation to his jungle self, he seemed to have become taller and at the same time brawnier and leaner, while his face had grown graver and harder.

"Excuse me for startling you," he said without the civilized smile he would have added [as his civilized self].[27]

The problem is that this transformation is too conscious, too studied. The appeal of Burroughs's Tarzan lies in the fact that he would never have taken note of the transformation, much less commented on it. Burroughs's Tarzan would have *undergone* the transformation without much transition at all. There are more subtle lapses, all of which detract in minor ways from the original character and the reasons for his extraordinary popularity. In a phrase, Tarzan is rendered too ordinary. This Tarzan, like so many others outside Burroughs's own writings, lacks depth. And yet *Tarzan and the Valley of Gold* remains the best of the pastiches, authorized or not.

Something of a maverick in the field of would-be successors to Burroughs is Philip Jose Farmer. In 1972 he published a putative biography of John Clayton. This authorized "biography" made no effort to carry on the Burroughs tradition but was described as the life of the man for whom John Clayton, Lord Greystoke, is the nom de plume. Significantly, Farmer describes this "changeling" as the "last hero in a world where the belief in heroes was dying out."[28] This volume is an honest tribute to Tarzan of the Apes, but it represents a footnote in the phenomenal popularity of the character.

Two other authorized pastiches bear mentioning here. The first is a film script by Gene Roddenberry, the other an authorized completion of a partial Burroughs manuscript. Neither works in terms of the unique depths of the character, and the latter is so far removed from Burroughs's writings as to be devastating as an introduction to the character.[29] Gene Roddenberry's script is an obvious attempt to bring out Burroughs's Tarzan with elements from that series and the author's Mars series, but yet again the mystique and depth of character are lost. *The Lost Adventure* makes essentially no effort to emulate the Burroughs style. It is further diminished by modern genre devices and (lack of) sensibilities. Both confirm the inability of any pastiche to match the "real thing" even as they indicate how deeply ingrained the character is in popular culture and how great the thirst for more tales of the ape-man.

Illegitimate Offspring

Given the popularity of Tarzan it is not surprising that there are a number of unauthorized pastiches. The most blatant of these is "The New Series" published in the mid-1960s by Gold Star Books. At best these unauthorized stories are an embarrassment to their authors. At worst they simply lift text right out of the original in outright plagiarism. Three of them[30] are representative in their lack of sophistication and failure to accommodate Burroughs's style. The dialogue is childish and stilted, with highly colloquial speech ascribed to Jane and Tarzan, where Burroughs had never allowed such a lapse. The patois of the time was always reserved to "stock" characters, usually as a kind of comic relief. Otherwise, Burroughs's characters always spoke perfect English.

One of the greatest offenses here is Jane's superiority in looking down on the "juvenile" behavior of her mate. Throughout these pastiches she is condescending to him, both directly and in the company of others. Tarzan himself is lacking in the essential dignity crucial to his universal and timeless appeal. He calls the apes silly, is sarcastic with his captors, and takes his status in the jungle very seriously. At one point he becomes frantic in an effort to escape captivity. He even stoops to lying to his mate, merely to preserve his self-image. It is difficult to imagine anyone more the antithesis of Tarzan of the Apes. As a potential introduction to the character these books would drive away the uninitiated reader before the originals could be sampled.

Because ERB, Inc., has guarded the Tarzan character so jealously, there are very few illegitimate pastiches extant. There are an array of tributes in the form of fan pastiches, all unpublished, but some making the rounds of serious fandom. Among these *Tarzan on Mars* may be the best known.[31] Despite the interesting concept of transporting Tarzan, Jane, and La of

Opar to Burroughs's Mars (Barsoom) where the three meet Tarzan's distant relation, John Carter (the human hero of the Mars series), the inconsistencies and lapses soon intrude and the narrative loses substantial momentum in the heavy-handed effort to correlate and explain the tenuous connections between the two series. Inevitably, the character of Tarzan deteriorates seriously yet again. In large part the problem is in Tarzan's consciousness of his status in both the jungle and society. Burroughs's sly juxtapositions between civilized mores and the ways of jungle society degenerate into obvious comparisons of little imagination.

Even less effective is another effort by Philip Jose Farmer.[32] An interesting curiosity, this effort is marred with a smug humor. Any advancement of Tarzan's popular image is lost as Farmer challenges Burroughs on issues there is every reason to believe the author comprehended full well but elected to ignore in the interests of characterization and rapid-paced narrative. The most accurate characterization of the effort is "cute," entirely out of keeping with the Tarzan canon. Farmer holds himself out as the American agent for Sherlock Holmes's Dr. Watson, Lord Greystoke, and similar literary figures and the disclaimer sets the tone: "All the characters in this book are real; any resemblance to fictional characters is purely coincidental."

All these pastiches have one element in common. They reflect the popular demand for more material of the Tarzan ilk. For more of this amalgam of feral man, noble savage, and Wild Man, it is necessary to turn to other characters.

Descendants in Spirit

Alter-egos Created by Edgar Rice Burroughs

When it comes to Tarzan, not even Burroughs was able to create a worthy descendant-successor. None of Burroughs's many heroes, however promising, possesses Tarzan's qualities; none is endowed with the truly unique mystique associated with the ape-man. The two who come closest are Nu[33] and Julian (The Red Hawk).[34]

Of the two, Nu comes closer to Tarzan, although Nu is not feral but a true wild-man, probably Cro-Magnon, precipitated by mysterious events to the Africa of John Clayton. Having died in an earthquake in his own true time, Nu is restored to life to resume his romance with the descendant (or reincarnation) of his lover. (In fact, this whole episode may be the vivid dream of the mildly disturbed American girl.) The result is Burroughs at his most intriguing and compelling. Nu might be considered a progenitor of Tarzan: "Nu, the son of Nu, his mighty muscles rolling beneath his smooth bronzed skin, moved silently through the jungle primeval. His handsome head with its shock of black hair, roughly cropped between sharpened stones, was high held, the delicate nostrils questioning each vagrant breeze for word of Oo, hunter of men."[35]

While the physical description recalls Tarzan (son of Kala), Nu is entirely atavistic without any strain of civilization to counterbalance the wildness and render the character a fascinating study in ambivalence. Still, as described by his lover, Nu does share something of the appeal of the ape-man: "And last night, in my dreams, I saw him again — alone and lonely, searching through a strange and hostile world to find and claim me."[36]

She is perceptive; Nu is not so simple as he might at first appear:

After these men and women [a party entertained by John and Jane Clayton] had eaten they came out and sat in the shadows before the entrance to their strange cave, and here again they laughed and chattered for all the world, thought Nu, like the ape-people; and yet, though it was different from the ways of his own people the troglodyte could not help but note within his own breast a strange yearning to take part in it — a longing for the company of these strange, new people.[37]

Thus, Burroughs suggests our longing for the wild has its counterpart in others' longing for civilization.

Like both Tarzan and Nu, Julian the Ninth is misplaced in his setting. Julian and his people are the feral remnants comprising a future human society on earth. Julian is marked by his awareness of all his incarnations — past and future. Julian is also tragic in a sense. He leads a doomed revolt against the resident aliens who have conquered the planet and is executed, but not before the next in his line has been sired. The Red Hawk is Julian the Twentieth, whose people have reverted to the Amerindian way of life.

Neither of these two Julians is developed to the same extent as Tarzan, but their respective ways of life make them kin to Nu and Tarzan. It is the lack of internalized ambivalence that renders Nu and the two Julians less interesting than Tarzan, although they stand well above the various pastiches of the ape-man.

Sheena, Queen of the Jungle

Sheena, the most widely recognized feral woman of modern genre fiction was introduced in Jumbo Comics in 1938.[38] In the 1950s she appeared on network television, portrayed by the statuesque Irish

McCalla. According to the television version, Sheena had lived among the beasts of the jungle since, as a child, she survived a plane crash. She was accompanied by the inevitable chimpanzee companion.[39] Sheena also reappeared in the mid-1980s in the midst of the superhero revival. Sheena, however, was less well received by critics and audiences than were her predecessors to the contemporary screen.

Some Lesser Descendants

Sheena was not alone among the distaff noble savages. Yet another indication of the power of this atavistic image is the array of tropical temptresses in a variety of media.[40] Along with Jane and Sheena, Nyoka of the Saturday serial and the array of Polynesians played by Dorothy Lamour are the most memorable. Additional male representatives are Roy Rockwood's Bomba and Otis Adlebert Kline's Jan.[41]

NOTES

1. De Camp's introduction to the Ace edition of *Conan of Cimmeria*, reprinted in the 1980s with a 1969 copyright.
2. Howard, de Camp, and Carter, *Conan of Cimmeria* (1969).
3. Ibid., 16, 73.
4. Ibid., 74.
5. Ibid., 49.
6. De Camp and Carter, *Conan the Barbarian* (1982), 49, 122.
7. Ibid., 96–97. See chapter 15.
8. William L. Chester, *Hawk of the Wilderness* (1935, 1936); *Kioga of the Wilderness* (1936); *One Against the Wilderness* (1937); *Kioga of the Unknown Land* (1938).
9. Chester, *Kioga of the Unknown Land* (1978 reprint), 10, 17, 41.
10. Ibid., 44.
11. Chester, *Hawk of the Wilderness*, 106.
12. Ibid., 76–77.
13. Ibid., 274.
14. Ibid., 275.
15. Nordhoff, and Hall, *Hurricane* (1935, 1936, renewed 1963), New York: Bantam Books, 1–2.
16. Ibid., 16.
17. Ibid., 20.
18. Ibid., 142.
19. Ibid., 189, 214.
20. See, e.g., Essoe, *Tarzan of the Movies* (1968).
21. "Short Waves," *The Gridley Wave* 25 (1968), 3.

22. Gardner, and Caruso, *The World's Greatest Athlete* (1973).

23. *Burroughs Bulletin* 33 (1974).

24. *The New Adventures of Tarzan*, 1967, Tarzana: ERB, Inc.

25. Leiber, *Tarzan and the Valley of God* (1966).

26. Ibid., 78.

27. Ibid., 141.

28. Farmer, *Tarzan Alive*, 30.

29. Undated script, *Tarzan,* with the notation, "National General Corporation/#1 Carthay Circle Plaza/Beverly Hills, California"; *Tarzan: The Lost Adventure,* attributed to Edgar Rice Burroughs and Joe R. Lansdale ("adapted and expanded") 1995, Milwaukie, Or.: Dark Horse Comics, Inc.

30. *Tarzan and the Cave City* (1964); *Tarzan and the Winged Invaders* (1965); *Tarzan and the Silver Globe* (1964); all by "Barton Werper."

31. Bloodstone, 1955, unpublished manuscript.

32. Watson, *The Adventures of the Peerless Peer*, in Farmer, (1974).

33. Burroughs, *The Eternal Lover*, 1914, New York: Frank A. Munsey Co. (*The Eternal Savage*, 1960, New York: Ace edition).

34. Burroughs, *The Moon Men* (1960). (Originally published in magazine serial format.)

35. Burroughs, *The Eternal Savage,* 5.

36. Ibid., 27.

37. Ibid., 31.

38. *The Gridley Wave* 20 (May 1996). See also Feret, *Lure of the Tropix: A Pictorial History of the Tropic Temptress in Films, Serials and Comics* (1984).

39. Grossman, *Saturday Morning TV* (1981).

40. Feret, *Lure of the Tropix.*

41. Rockwood, *Bomba the Jungle Boy: Or the Old Naturalist's Secret* (1926); Kline, *Jan of the Jungle* (1931). This latter volume is frankly advertised as "Great jungle adventure in the Edgar Rice Burroughs tradition" on the cover.

15

The Ape-Man's Modern Descendants and Remote Relatives

Although they are transcended by a certain evocative quality associated with each representative of this emergent archetype, the continuous evolving Wild Man lineage is characterized by the four features listed below:

1. This Wild Man, usually a changeling, is of two worlds, literally, figuratively, or both.
2. He is something of a loner and estranged from at least one if not both of those worlds because of some degree of ambivalence regarding his proper place.
3. Despite the estrangement, he functions with extraordinary effectiveness in one if not both of his worlds.
4. His presence manifests a commentary on contemporaneous human civilization, at once admiring and criticizing it through comparisons to the alien society that is his.

If there is a fifth feature, it is that he is in a very affective sense us.

IN THE CINEMA

Because the concept of feral man is probably best known to the public through the cinematic Tarzan, films dealing with concepts of this version of the Wild Man are an important source of descendants. In the course of such films, the definition broadens until the larger genres of science fiction and fantasy become the vehicle. With this broadening, the transcendent Wild Man includes a greater diversity and sophistication. It is possible to proceed from Tarzan to *Star Trek*'s Mr. Spock with no lapse in logic.

The Savage

Among the relatively early cinematic Wild Men of this ilk is Jim Ahern, protagonist of the 1950s film, *The Savage*.[1] As a boy, Ahern, the sole survivor of a Crow massacre of his wagon train, is adopted by Sioux and raised as War Bonnet, son of the tribal chief.

Upon entering manhood, War Bonnet becomes the object of constant criticism as confrontation with the invading Caucasians becomes imminent. No one is certain which side he will take. At first War Bonnet's loyalty is unwavering: He is without question Sioux. His foster father, however, is wiser in human nature and asks only that the young man not disgrace his name. War Bonnet's reply is as profound as it is sincere — and has a familiar ring to it: "Is it the pigment of a man's skin which makes him Miniconju? The color of his eyes? Neither of these things. No it is the beating within his body."[2]

Unfortunately, War Bonnet, as the offspring of both societies, is dispatched to the nearby fort to discover when the invaders plan to remove the Sioux to reservations. On the way, he rescues a party of soldiers from marauding Crow. War Bonnet's subsequent assimilation into his former society meets with mixed success. His unswerving loyalty to the Sioux wavers ever so slightly. Then intervening events and misunderstandings result in the death at the hands of soldiers of a Sioux girl close to War Bonnet. This event drives the embittered War Bonnet back to the Indians as his true people. They ask him to make their case to the soldiers. Seeking to find his place between cultures War Bonnet is, once again, put to the test of his loyalty when he is charged with the mission of leading the soldiers into a Sioux ambush.

If War Bonnet starts out without ambivalence, the fact that the soldiers undertake the protection of a wagon train shakes his resolve and once again calls into question his ultimate loyalty. In the end War Bonnet warns the soldiers that he has seen Indians ahead and thus succeeds in averting the ambush without betraying the Sioux. Knowing his life is forfeit for his act of apparent disloyalty, War Bonnet returns to his Sioux family to plead for peace — for their sake.

His father listens to his words but is compelled by tribal law to strike him down. By accident or design the thrust of the spear is not fatal, and the Sioux return the wounded War Bonnet/Jim Ahern to the fort where, presumably, he will make a life with the woman he had begun to love during his earlier stay in the fort.

Simple in the bare-bones outline, this little-known film is an eloquent statement of the ambivalence of the changeling. It its day, *The Savage* was a prescient commentary on inter-racial relationships, in comparing so-called civilization to a primitive way of life, to the *former's* detriment. War Bonnet himself is in a sense feral and is very much the Wild Man who leads not one but two peoples to a better understanding of their mutual destinies while revealing his own true nature.

Apache

More akin to Tarzan, Massai of the film *Apache*[3] finds himself trying to make his way in the emerging world of the successful invaders of Indian lands. Massai is represented as the last Apache warrior. He escapes as he is being transported with Geronimo to Florida.

As Massai makes the long trek home he finds himself in the bustling city of St. Louis. The strange ways of the people he encounters there are a source of bemusement. There is, however, no loss of this man's inherent dignity as he observes these strangers in silence. It is apparent Massai has no interest in becoming part of any such "civilized" culture. But then circumstances force this loner to take refuge in a farmer's barn.

The farmer, Massai discovers to his astonishment, is Cherokee. The man lives in a "white man's house" and has accepted the ways of the invaders, with whom he lives side by side. Preferring to return to his own people, Massai declines the invitation to remain but does reluctantly accept some seed corn. Upon his return, Massai finds his people subjugated by the military. He plans to preach what the Cherokee farmer taught him, but he is recaptured. The heroine, who was to have been Massai's wife, has the courage to advise the soldiers of Massai's "reformed" intentions, but they remain unconvinced.

Another escape ensues, but Massai is now estranged from the Apache world as well as that of the invaders. He is deeply embittered, and the woman calls him a dying wolf biting at his own wounds, fighting only for self without regard for his devastated people. He actively seeks vengeance and his own death.

Ultimately, the two become man and wife and together continue to elude the parties of both races searching for Massai. He makes a true home for his family. He has become a legend among his people and even the military would call off the private war. But his small farm is discovered just as the birth of his child is imminent. Against the urging of his wife to escape alone, Massai declares that only a warrior chooses his place to die and observes he is no longer a warrior. At her urging he goes to face his enemies to win his warrior's death, that she might sing of it to their child. After a prolonged battle in and around the thriving field of corn, the attack is terminated with the first cry of the newborn. Massai is, after all, the first Apache to take up farming. In this alteration hope for the future is seen.

For all the contemporary political incorrectness of this film, Massai remains the noble savage, if not truly feral man caught between two worlds. He is the archetypal hero and would-be savior of his own world but must accept the ways of an alien world to survive. As he boldly attempts to establish for himself the best of both worlds while remaining true to his heritage, Massai takes his place as a legitimate member of the modern family of Wild Man.

First Blood

Among the most recent cinematic Wild Men is more rebel than most. He is John Rambo of *First Blood* and its first sequel.[4] Rambo has been made an alien by his own society, perversely resulting from his medal-winning efforts as a Green Beret in Vietnam. In the first film a minor defiance of local law enforcement escalates to an extraordinarily violent escape, leading in turn to a prolonged manhunt, with Rambo becoming something of a noble savage along the way. The pointed commentary on civilization remains implicit throughout, approximating the predominant theme of the underlying novel. In the end Rambo collapses in the presence of his empathetic former commander. The original provocations of his untamed violence are expressed in an inarticulate but profoundly moving soliloquy. Rambo's pleas are to the civilization which has renounced him and his lost colleagues-at-arms. He speaks not just for himself in the implied demand that we reevaluate our own attitudes.

Rambo is even more akin to feral man in the sequel. With little preliminary Rambo is released from prison to return to the jungle wilderness in which his skills had been honed. This film expands upon the abilities of a jungle guerrilla fighter. This time, however, Rambo is not commenting on our civilization but has been elevated to our surrogate in belatedly defeating a more traditional military foe after our own society had been precluded from doing so. Along the way, he defeats representatives of those who precluded that victory. It is significant that Rambo's code name is Lone Wolf. Virtually invincible in his aloofness, this archetypal hero and savior is superior in every way to the more traditional forces of both sides of the physical conflict.

Notwithstanding critical disdain, the success of these two films is eloquent testimony to the perpetual human need for such superhuman heroes, especially those of complex character, including deep ambivalence about their proper place between civilization and the wild.

Planet of the Apes

Cinematic science fiction has made the point that just who is the alien depends on whose *terra* is the setting for the action with Taylor, who is thrust into the alien role upon landing on *Planet of the Apes*.[5] Here there are some interesting juxtapositions: The noble savage of the piece is a living commentary on the civilization of the resident apes, while representing to them a dangerously intelligent beast. At the same time both the apes and Taylor are a less-than-flattering commentary on our own civilization. Taylor eventually assumes a feral way of life in an effort to improve the lot of what proves to be a feral community of humans, descendants of our own future mistakes.

In one telling scene the influential scientist, Zaius, explains to the defeated beast/savage why he must be eliminated for the sake of ape civilization. As is often the case with the Wild Man, the danger is in his ability to undermine currently accepted civilization. It is his very intelligence that dooms Taylor, condemned for the hazard he poses to the culture. Taylor escapes with his recently acquired mate and sets out for what promises to be a solitary existence that could ultimately restore humanity to supremacy. Initially an arrogant representative of humanity, Taylor may become the archetypal hero-savior.

IN LITERATURE

That feral individuals remain in the province of "literature" as opposed to genre fiction is evidenced in the critical reception of a child's story of a feral Eskimo girl.[6] The girl, Miyax or "Julie," becomes lost in an attempt to escape a grossly unhappy situation and, at age thirteen, turns to wolves for help as once her father had. Julie studies the pack to discover the signs of good will and friendship and learns to "talk" to the animals. Eventually she becomes one of the pack.

Seeking a friend in San Francisco, Julie makes her trek in the company of wolves part of the time, and they actually save her from an attacking grizzly. Later, when hunters slay the leader of the pack, Julie protects a second member and, in doing so, comes to realize she cannot accept civilization. She turns back, however, to the old Eskimo ways rather than to a truly feral life among the wolves.

More distant relatives are to be found in mainstream fiction, represented here by Ayn Rand and a retelling of an ancient legend.

Ayn Rand's Roark

We find Roark in a Connecticut quarry: "He stood on the hot stone in the sun. His face was scorched to bronze. . . . The quarry rose about him in flat shelves breaking against one another. It was a world without curves, grass or soil, a simplified world of stone planes, sharp edges, and angles."[7]

Roark is descibed in terms reminiscent of many a wild man: He is first seen naked on the edge of a cliff high over a lake. His body is described in "long straight lines and angles, each curve broken into planes," not unlike the quarry. Roark is described as "moving with the soundless tension, the control, the precision of a cat" and "relaxed, like a cat, in shapeless ease, as if his body held no single solid bone." People stared at Roark as he passed "with sudden resentment. They could give no reason for it; it was an instinct his presence awakened in most people."[8] Roark does not care what people think of him, although they are irresistibly drawn to his talents. Keating asks him why he cannot be human, simple, and natural, to which Roark replies, "But I am."[9]

Working for a man he despises, Roark is in essence a wild man or noble savage held captive, that is, apart from his wildness. He went to work in "the prescribed pearl-grey smock like a prison uniform on his body." In this job "[h]e had to obey and draw the lines as instructed. It hurt him so much." "But he knew this would not last — he had to wait — it was his only assignment, to wait —" for freedom, of course. Roark, like other wild men, has few, select friends.[10] A young photographer who loves Roark's work catches the man looking at a house he designed and built on New York's East River. Roark

Was leaning back, . . . looking up at the building. It was an accidental, unconscious moment. The young photographer glanced at Roark's face — and thought of something that had puzzled him for a long time: he had always wondered why the sensations . . . in dreams were so much more intense than anything one could experience in waking reality . . . and what was that extra quality which could never be recaptured afterward; the quality of what he felt when he walked down a path through tangled green leaves in a dream, in an air full of expectation, of causeless, utter rapture — and when he awakened he could not explain it, it had just been a path through some woods. He thought of that because he saw that extra quality for the first time in waking existence, he saw it in Roark's face lifted to the building.[11]

Roark is a wild man, even Wild Man; he is of the lineage of Tarzan, the noble savage. Theirs is a strength, a wildness, an inability to be purely wild or purely civilized "made visible" for us.

"Finn McCool"

This recent novel is a retelling of the Irish saga of an early wild man. His feral bearing is emphasized with a description of Finn as a rawboned youth of massive frame, improbably broad shoulders, and a disconcerting smile marked by a "feral baring of teeth" upon confronting a stranger. His hair is described as molten silver and his cheekbones as boulders. He wore a "huge, rough mantle of wild-animal skins crudely stitched together with sinew." Beneath his gaze flickers "something as hard and cold as iron." His lurking quality can disappear into good humor.[12]

Finn's mother bore him in a bog while fleeing enemies of his father. Finn is raised by a druid and a wise woman. Perhaps learning the magic of the druid, he can disguise himself as a deer and holds certain powers. Upon Finn's initial approach to the king he will serve, the king notes his boldness and difference with the smell of the wild about him — the king likes wild things. In personality, Finn is able to switch in a flash from wild and dangerous to amicable. Although well received at the High King's Tara, Finn harbors a "shy, wild boy" within who "wanted to run off into the forest and bathe himself in silence."[13]

Like so many wild men, Finn at times prefers to pass the night in the forest and longs to remain there; his "body tingled with the elation of being one with the forest again. The fragrance of the earth and leaf mould was in his nostrils. He wanted to sleep pillowed on a tree root and feel the night alive around him." And yet he is poetic. His favorite music is the sounds of wild fowl and bird and of cattle. His poetry is of place and comes to him when "out under the sky. In the silences," he hears it.[14]

Finn is not readily "knowable" by others. With his quiet and poetry, one observes "he goes off someplace inside his head when he's composing or reciting poetry, someplace we aren't allowed to follow him." This distancing occurs in the wild, too, when "we're in a forest, or on a hilltop" when he seems to glow, as if "he's come home again. He gets a look about him I never see on his face when he's under roof." In short Finn is wilder than his comrades. He does not subdue the wildness when confronting the king, notably over treating Finn's injured hound. Inclined to withhold his physician's services the king feels for the first time the full force of Finn's will. "Finn's posture did not change, nor did he move a muscle in his face. Nothing changed. Yet a Thing looked out of his eyes that had not been there before, a creature more feral than a wolf, more terrible than a storm. A creature well beyond imagining, an elemental force, a power capable of destroying everything in its path if opposed."[15]

Finn marries and loses a woman whose own ancestors are among the fairy folk. Upon encountering what appears to be her spirit, he tells her that he grew up wild, even the king had once with some accuracy described him as his wolf cub. He had run "like a wolf cub, untaught and unfettered" and left the women who reared him when he could survive alone. Even the defenses Finn designs for Tara are the patterns of "a mind accustomed to the wild ways of the wilderness, to tangled forests and secretive glens."[16]

Upon being personally challenged by his lifelong rival, Finn's face changes subtly, "as if the bones beneath the skin shifted ever so slightly." As the other watched, "they realigned themselves" so the "face was no longer that of Finn McCool. The hot, bright eyes looking out of it were not even human." And the challenger knew this "wild creature . . . would kill him without conscience and enjoy the hot spurting of his blood."[17]

This is precisely the haunting and timeless appeal of the wild man. He is *half* tamed. Neither fully wild nor completely civilized, he is a metaphor for our own ambivalence about the proper balance between civilization and wildness.

SCIENCE FICTION AND FANTASY

In Popular Fiction

Although the foregoing suggests just how pervasive the wild man is in popular fiction, most contemporary tales of feral individuals are found in works of science fiction and fantasy. Many of these tales involve feral populations of humans rather than isolated individuals. The settings reflect a future population seeking to survive following some natural or manufactured catastrophe that has wiped out civilization as we know it. A particularly readable entry in this subgenre tells of a loner who happens to be a graduate ecologist. Eventually he founds a new civilization in the company of other survivors he gradually gathers about him.[18] The novel is a fascinating contemporary piece incorporating elements reminiscent of such classics as *Swiss Family Robinson* and *Robinson Crusoe* to tremendous effect.

Then there are the works of alternative history. One of the best in this subgenre is Orson Scott Card's *Hatrack River*.[19] Card's Red Prophet is a version of "Ta-Kumsaw." While there is nothing novel in the impression of an Indian as one with the earth, authors like Card can embue the image with fresh vitality and a sense of reality.

Like his ancestors', Ta-Kumsaw's "harmony with the woodland was complete. He did not have to think about where he placed his feet; he knew that the twigs under his feet would soften and bend, the leaves would moisten and not rustle, the branches he brushed aside would go back quick to their right place and leave no sign he passed."[20]

Card calls the relationship with the earth "green music." What Ta-Kumsaw is perceiving is "the green life of the woodland" and beyond into "the black whirlpool sucking him downward, stronger, faster, toward the place where the living green was torn open to let a murder through." The murder was that of his father, who would be given back to the land — literally.[21]

As a child Ta-Kumsaw's brother was witness to this murder, which turns "Lolla-Wossiky" to liquor, which causes the resultant "black noise" to fade "[i]nto beautiful green silence." But that green constantly disappears, fading into the black. This green is the Indian land sense, particularly vivid in Lolla-Wossiky when he can banish the black. His loss of the green leaves him like the Caucasian: "Cut off from the land. Ground crunching underfoot. Branches snagging. Roots tripping. Animals running away."[22]

Ironically, the Caucasian protagonist of the novel restores the green music to Lolla-Wossiky. It is a profound moment "as the land returned to him the way that it had been before. . . . [H]e had forgotten how strong it was, to see in all directions, hear the breath of every animal, smell the scent of every plant." In that moment, Lolla-Wossiky discovers that the black noise was not a thing but an emptiness at the "place where the land ended, and the emptiness began."[23]

Also close to the concept of the Wild Man are the heroes who are the last of their kind, faced with an evolving new species certain to be a (lesser) replacement. In this context the most significant representative is Michael Moorcock's eternal champion, especially in the Corum incarnation.[24]

Corum

Corum is the last of his race, the Vadhagh, succumbing to the increased numbers and brutality of the Mabden, quite obviously Homo sapiens.

Thrust brutally into the Mabden world, Corum is forcibly taught first sorrow, then fear and rage. He thus learns for the first time how to kill and to be cruel. He gains cunning and the lust for revenge.[25] While superior to the invading alien race, Corum becomes both ambivalent and beastlike in his response to this enemy, among whom he is destined to thrive:

For all that the Mabden had saved his life and seemed both cheerful and courteous, Corum had learned, as an animal learns, that the Mabden were his enemies. He glowered at the Mabden.

"What have you saved me *for,* Mabden?"[26]

While Corum is a sophisticated creature of a superior world, there are clear parallels to the tales of the patriarch, Tarzan. Beyond the bestiality human cruelty has imposed upon him, Corum's life comes to turn about his love for a woman of the world otherwise alien to him. She is Rhalina, a heroine who in many ways is comparable to Jane Porter Clayton. Rhalina holds a metaphorical mirror before Corum in which he can see himself as does the alien world. More importantly to him, she makes his acceptance in that world a worthy goal: "Corum knew he was mad, in Vadhagh terms. But he supposed that he was sane enough in Mabden terms. And this was, after all, now a Mabden world. He must learn to accept its peculiar disorders as the norm, if he were going to survive. And there were many reasons why he wished to survive, Rhalina not the least among them."[27]

Note that with both Corum and Tarzan before him, the choice of this lineage of Wild Man falls between two worlds, never quite in either. Corum cannot return to his world, lost to our civilization. His choice is between our civilization and death. And like Tarzan before him, Corum has the capacity to horrify those who care for him because of his alienation from their familiar, comfortable society:[28] To Rhalina Corum is at once frighteningly "other" and more than human. He is simultaneously very much a tragic figure, this time overtly. Corum is then the archetypal hero-savior, albeit a reluctant one.[29]

Above all, Corum is ambivalent in the role Fate has thrust upon him:

He was unlike others of his race, though he had the same tall beauty of form. . . . [H]e was Corum Jhaelen Isrei, . . . who desired nothing but peace but could not trust the peace he had.

"Chaos only seems more powerful because it is aggressive and willing to use any means to gain its end. Law endures. Make no mistake, I do not like the role in which Fate has cast me — I would that someone else had my burden — but the power of Law must be preserved if possible."[30]

Androids

Do Androids Dream of Electric Sheep?[31] presents rather a cerebral version of the Wild Man, who may be found in the female android, Rachel Rosen, or, perhaps, in the protagonist, Rick Deckard, who may himself be an android. Neither he nor the reader will ever be sure which world is Deckard's. Both characters exist in the milieu of a future human community that has degraded not quite to the point of being feral. Both offer an implicit commentary on that society. In the original film, the Wild Man could very well be the leader of the rogue "replicants."

Dolphins

On a very different level Roy Meyers has created a feral boy, superficially not unlike Tarzan; one who is nurtured by dolphins.[32] Given the combined fascination of the ageless lore of dolphins and of feral man, the series is a disappointment. All the promising elements of atavastic adventure are there, but neither the principal character nor the events are developed satisfactorily. It may be that the author aimed his material to a young audience without any expectation of a deep response transcending that primary readership.

N'Chaka

Closer to the image of Tarzan and Corum is Leigh Brackett's Eric John Stark — N'Chaka, Man-Without-a-Tribe. The details of Stark's early biography are not provided. Instead, hints of it are from time to time slipped into the narrative. Stark, a Terran, was born in a mining colony on Mercury. Upon being orphaned on this inhospitable planet, he was fostered by a tribe of sub-human aboriginals. Humans slaughter the tribe and cage N'Chaka, "a snarling curiosity. He survived to fight for survival as a man."[33]

Stark, like his literary ancestors, is an alien among his own species: "I'd lost all my human origins, of course, and the humans I'd met had given me little cause to love them. Ashton . . . tamed me. . . . Yes, he [gave] me much more than just my life." Also like those ancestors, Stark is given to introspection: "A wolf's-head, a man without a tribe. I was raised by animals, Jerann. That is why I seem like one."[34]

Stark seldom reveals much of himself, but in repose much can be observed:

Control. That was the strength one felt in the Dark Man. Strength that went beyond the physical. . . . Stark's face fascinated Tuchvar. . . . He thought it was beautiful in its own way. Subtly alien. Brooding, black-browed, with a structure that might have been hammered out of old iron. A warrior's face, scarred by old battles. A killer's face, but without cruelty, and when he smiled it was like sunlight breaking through clouds. Now, in the unguarded innocence of sleep, Tuchvar saw something there that he had never noticed before. It was sadness. In his dreams, it seemed, the Dark Man remembered lost things and mourned them.[35]

Like Conan, Stark becomes a mercenary. He is intelligent in a feral sort of way. He is human enough to possess a bitter sense of humor.[36] But Stark is a little too human to qualify as a noble savage. He has a transcendent need for freedom and yet, with a characteristic ambivalent stoicism, will accept its loss: "He recognized the inevitable. He was used to inevitables — hunger, pain, loneliness, the emptiness of dreams. He had accepted a lot of them in his time. Yet he made no move to surrender. He looked out at the desert and the night sky, and his eyes blazed, the desperate, strangely beautiful eyes of a creature very close to the roots of life, something less and more than man."[37]

This tradition of ambivalence, so well phrased above, is very much a part of Stark. He uses his own animal powers both defensively and offensively against other animals. And yet, the combination of abilities can be something of a mixed blessing:

Stark's human reason told him that these monsters were no more than lumps of eroded stone. His mind knew that. His primitive gut said otherwise. And his animal senses told him that other beings not of stone were close by.[38]

Of course, it began with his first people, who shared all with N'Chaka — They taught him love, and patience, how to hunt the great rock-lizard, how to suffer, how to survive. He remembered their faces, wrinkled, snouted, toothed. Beautiful faces to him, beautiful and wise with the wisdom of first beginnings. His people. Always his people. Always his people, his only people. And yet they named him Man-Without-a-Tribe.[39]

As a result of his feral upbringing, Stark-N'Chaka has supranatural senses and bestial patience and cunning;[40] he hates captivity and will fight for his life with irrational force. "Stark's senses had developed in a strange school, and the thin veneer of civilization he affected had not dulled them."[41] Stark, unlike Tarzan, feels fear and, more, is not in the slightest shamed by it. His philosophy holds that "The man who doesn't fear, doesn't live long."[42]

Stark, like Tarzan, is human and more. But he is not comfortable among humans, of his own Earth or elsewhere, although he does enjoy human pleasures. He never finds, however, the soul-mate of either a Tarzan or a Corum. In fact, Stark never makes any firm commitment to anyone or anything. It is Stark, though, who maintains a certain lightness of spirit, even when pondering his internal alienation. His two halves are often engaged in dialogue: "It was not a good land. The primitive in him sensed evil there like a sickness. It wanted to turn tail and go shivering and howling back to the smoky warmth . . . and the safety of walls. The reasoning man in him agreed, but kept moving forward nevertheless."[43]

Battle will bring out N'chaka, but Stark is in charge at its conclusion. His animal side makes Stark an unpopular companion among some, who call him "the wild man" with derision, not respect. Indeed, he does revert to N'Chaka easily: "Eyes opened and stared at him, and Thord could not repress a slight shiver. . . . He had seen exactly the same gaze in a big snow-cat caught in a trap, and he felt suddenly that it was not a man he spoke to, but a predatory beast."[44]

The reaction of a seeress is understandable confusion. She asks Stark to tell her who he is — "A lesser gift even than mine could sense that you're different. Inside, I mean. There's a stillness, something I can't touch."[45] His mentor, Ashton, refers to Stark as "a tiger wandering from one kill to the next" and urges him to make a [hu]man of himself. Had Stark matured as did both Tarzan and Corum, he could have been a worthy successor to the ape-man.

Vampire

Perhaps unique among Wild Man as fantastic hero is Saint-Germain, Prinz Ragoczy of Transylvania, Chelsea Quinn Yarbro's vampire.[46] Saint-Germain is introduced as a mysterious gentleman of no more than forty-five years. All the ladies are fascinated with this composer and performer on the violin and harpsichord. In fact, Saint-Germain was already two thousand years old in the days of the Roman Empire.[47] Among his other talents, Saint-Germain speaks at least five languages and possesses extraordinary strength. But he has been a slave and has suffered great loss during his prolonged "life."

Nevertheless, Saint-Germain is described as elegant and delightful. He moves silently, "his well-knit body moving with fluid grace remarkable in a man of his age."[48] He has intense, dark eyes others find fascinating and compelling. Whatever the society in which he may find himself,

Saint-Germain is extremely urbane and witty. He is also clever and compassionate, as when a friend discovers the lack of reflection. Saint-Germain explains the phenomenon away as a matter of the angle of observation. Clearly he is protecting his friend, not himself.

Saint-Germain is a man of honor, who is taken aback when that trait is not accepted at face value. But he is also a man of bleak thoughts and untouchable remoteness. He is no longer able to weep. "For him, all pain, all anguish, was inward, and there was no release in tears. There is a bitterness about him: 'Do you imagine . . . that I could reveal my true nature and be treated with anything other than repugnance and detestation?' His voice held more suffering than anger."[49]

Saint-Germain's physical attributes are in keeping with his unique nature; he has a feral quality: "His movement through the shadowed night was fluid, was powerful, was beautiful. His grace was not the grace of a dancer, whose splendid ease is born of meticulous, disciplined years, but rather it was his natural condition, an aspect of self as much as his musical voice and arresting eyes."[50]

Not only his enemies, but also his friends recognize the basic forces in the man — formidable and compelling. But he has, of course, paid a great price for his attributes and it has estranged him from most of humankind, even though the rare individual is drawn to him in friendship. But even in such circumstances or when Saint-Germain elicits admiration in others, he himself feels profoundly alien. In this man, as in Tarzan and others, the conflict between human-ness and another inherent nature represents both the best and the worst characteristics.

On occasion, Saint-Germain enjoys close friendships with men, but his alienation is eased by his relationships with the few women in his life. Those women who, throughout the series, accept him as friend or lover or father all come quickly to accept his nature. Their respective commitments to him are as deep as his to them. Collectively, they represent Corum's Rhalina or Tarzan's Jane. They are Saint-Germain's link to his human side, which remains very much alive.

On Television

If nothing else television is the medium most nearly ubiquitous in this country and around the world. Television has produced its share of representatives of the Wild Man, including the Tarzan series themselves. As early as 1957 there was a syndicated series called *The Adventures of a Jungle Boy*.[51] A more recent television movie, ABC's *Stalk the Wild Child*,[52] seriously treated the nature of feral children. The protagonist, dubbed "Feral Cal," is a boy of perhaps eleven years found living among a pack of wild dogs in California. He is found by hunters and taken to a hospital.

Mute, Cal has forgotten his association with humans and exhibits characteristics reportedly common to feral individuals. Cal is extremely fearful of people and capable of neither laughter nor tears. A team of psychologists (Dr. Hazard and Maggie) seeks to guide Cal into his human heritage, although they are well aware time is running out, if only because the boy's "teachable moments" will soon be lost. Eventually Dr. Hazard takes him in as a foster child.

Once in a homelike setting, Cal's progress is superficially rapid, but for speech. His progress is marred by his rebellion. His longing to return to the wild makes him violent as well as uncooperative. It is not until Maggie threatens to leave that he utters his first words and then proceeds to writing and to using tools and progresses into relatively normal teens. He even graduates from college.

Then, with Maggie, Cal attends a lecture which is illustrated with films of his early reunion with his species. Members of the audience, even Maggie, are amused by his behavior. Shamed by their reaction, Cal leaves precipitously. Maggie follows and finds him tearing apart the computer room in which his records are maintained. She and Dr. Hazard explain to Cal that parents do laugh at their children and that people often laugh defensively, but the youth is unconvinced. At length Dr. Hazard assures Cal that he is like a son, but Cal retorts with considerable justice, "Then my name should have been the same as yours."

Cal runs away, completely disillusioned. It is now that he meets a publisher who wisely addresses him as Mr. "Farrell" and uses his given name only with permission. Cal responds to the respect and listens to the man's assurance that his immense inner journey from primitive to civilized man would contribute to our knowledge of ourselves. Still suspicious, Cal asks why the publisher would do anything for him, to which the other asserts that he likes the boy while candidly admitting he is looking for a best-seller.

Cal enters his home, the forest, to consider, and he begins to set down his experiences and reactions. He is joined by an editor, Andrea, who describes his experience as simultaneously terrifying, beautiful, and tremendous. Cal is smitten and reassured by her professional interest — until he sees the final product, entitled *California Wild*, with a lurid cover illustration. Despite Andrea's reassurances Cal is devastated and runs off into the forest. Andrea follows until she is frightened by the sound of dogs and turns back.

Cal continues his wandering and one day comes upon a tattered kite with a bell, the sight and sound of which recall moments of his past and the incident that had resulted in his prolonged isolation. He is restored to his own family and, through them and his love for Andrea, renews his commitment to human society, despite the continuing enticement of the forest.

In 1983 the prime-time series, *Tales of the Gold Monkey,* a mild spoof of exotic adventures set in the 1930s, took on the feral child.[53] This teenaged boy lived among apes on the otherwise uninhabited windward side of a remote island. While the villains are attempting to capture him, the boy has mistakenly concluded that the hero and heroine are his parents. Remaining nameless throughout, the boy has remnants of language, notably "mama," "papa," and "dog," but his behavior patterns are ape.

Eventually the decision is made to allow the boy to make the choice of life he is to pursue. Although he is terrified and snarls at his father's approach, the boy calms when shown a copy of a photograph in his possession, showing him with his parents. While this story adds little to the family history of the Wild Man as feral man, it is suggestive of the persistent psychological appeal of this manifestation, even without the romantic conceits of genre fiction.

Two television series themselves centered on feral individuals, broadly defined.[54] One, *Lucan,* featured "Cal" at about twenty. Although acclimated to civilization, Lucan is still ill at ease in cities. The series follows Lucan's search for his parents and his own identity. The other, *The Quest,* was not centered on truly feral men but on two Caucasian brothers raised by Cheyenne. They are living in a Caucasian world but tend to be more trusting of Indians. Significantly, the Indian name of one is Two Persons.

The original *Kung Fu* series[55] represents an interesting bridge between the feral man and noble savages and a branch of the family Wild Man that is truly unique to the latter half of this century. Caine is hardly feral, but he is portrayed as a variant of the noble savages. He is alien in two senses of the word. Chinese-American in the mid-1800s, Caine wanders the West, accepted neither by the Chinese nor Americans. Caine necessarily remains aloof and alone. Yet he represents a constant commentary both expressed and implicit on all he surveys and is forever making right the wrongs he encounters as he searches for his identity.

Television's science fiction has provided members of the Wild Man lineage, literal aliens who are at most only partially *Homo sapiens.* Certainly the best-known, most phenomenally popular of these alien Wild Men is *Star Trek*'s Mr. Spock, yet another indication of how deeply the concept affects us. That the character has had its influence on the one most responsible for creating it is manifest in Leonard Nimoy's two books, written some twenty years apart[56] and themselves indicative of Spock's, Nimoy's, and our own profound ambivalence.

Gene Roddenbury's Spock stood out from the beginning. That the character is of the lineage addressed here is immediately apparent in Roddenbury's response to network reaction to Spock: "My own idea . . . was, in a very real sense, we are all aliens on a strange planet. We spend most of our lives reaching out and trying to communicate. . . . [A]nd this is exactly what Spock is trying to do."[57]

Over time the original series reveals that Spock's parents are Vulcan and Terran, ordinarily incompatible stock. Spock is of both worlds, but he considers himself Vulcan, a race most obviously reflected in his stoicism. But all of Spock's senses are better developed than those of humans. Beyond the comparably increased fortitude Spock can control his own consciousness. "Spock's stoic temperament, his refusal to say anything or do anything not based solely on logic is also a reflection of his Vulcan heritage. Complete adherence to logic is the primary motivating factor in the Vulcan mental process. Of necessity, complete suppression of emotions is required, lest logic be influenced in any way."[58]

Were this the extent of it, Spock would be speaking of our relentlessly civilized natures bound to the material realm. He would be a true superman and substantially less interesting. Hence, his fatal flaw (and his strength) are combined in the fact that at his core, Spock's logic and his emotions conflict. "The result is a continual struggle within himself to suppress his feelings. But his Vulcan side is normally in control. Conditioned since childhood not only to deny but also to be ashamed of emotion, Spock thrusts feelings aside and finds a 'logical' rationalization to explain it."[59]

Since he is out of place on both Vulcan and Earth, a thing his Vulcan father finds inexplicable, Spock entered Federation service. With this choice comes estrangement from his father and loneliness. Spock is the sole Vulcan on the Enterprise and is capable of minimal social intercourse. But over time (and several feature films produced long after cancellation of the original series, which remains in syndication to this day) Spock's abilities, along with his complex personality are gradually evoked, making him a valued friend, as well as commentator on the rather benighted civilization known as humanity.

It can come as no surprise that, most of all, Spock is "biologically, emotionally and even intellectually a half-breed."[60] He "is the alien, the outsider, the stranger" and the "hero of an ancient . . . fantasy — the dream and the reality of being the different one, the lonely one, yet also the perceptive one, the one who understands." "Science fiction is full of [his] forerunners . . . , [but] never one more perceptively tailored to speak exactly to our most urgent problems: loneliness, alienation, changing and clashing values, the need to find home and friendship among strangers, the need to master our world and ourselves — especially ourselves."[61]

Aside from Mr. Spock, three aliens stand out: Mark of *Man from Atlantis,* Questor of *The Questor Tapes,* and Hawk, a secondary lead introduced in the final episodes of *Buck Rogers in the Twentyfirst Century.* All three showed tremendous promise as modern standard-bearers for the family Wild Man, but the promise was ultimately lost either in execution (Mark and Hawk) or because the vehicle was not sufficiently well received to attain series status (Questor). Mark is essentially a creature of the sea.[62] Questor is neither human nor strictly alien, but an android.[63] Questor is assembled according to instructions left by a scientist who has mysteriously disappeared. The assembly is incomplete, leaving Questor to finish the task as best he can, leaving him entirely human in appearance but marked by a few conversational quirks and a certain mechanical quality to his movements, both of which gradually disappear. The pilot is Questor's search for his identity in the company of his mentor, a cybernetics engineer.

When the film closes, Questor has been reassured of his ability to serve so long as Jerry continues as his mentor in matters of human emotion. But how this unique member of the Wild Man lineage, neither truly alien nor quite human, will grow intellectually and emotionally, remains a mystery as no series followed. Data, the android of *Star Trek: The Next Generation,* does not reflect any kind of ambivalence, although he, too, offers commentary on civilization. Another remote descendant of this lineage is Duncan MacLeod of *Highlander: the Series,* although he, too, lacks ambivalence for the most part and offers little commentary on civilization despite his alien nature and fondness for humanity. He may or may not be a changeling; the series has never identified his true nature, although MacLeod, upon his first revitalization, once asked his father who he really is.

Hawk[64] is from an alien world of our distant future, but his roots, although Terran, are eventually revealed to be nonhuman. Buck Rogers, commissioned to bring this outlaw to justice, discovers him to be part avian, a descendant of a bird people of earth's Polynesia. Here there are elements from a variety of estranged noble savages, increasingly apparent when Hawk is taken prisoner after an air battle in which his mate, Cory, is gravely injured. The loyalty and the profound need for freedom are both featured, as is the tragic history of Hawk's people.

As the episodes reach their climax, the combination takes the form of the need for Hawk and Buck Rogers to complement each other against attacks from others. But they come again into conflict after the death of Cory, when Rogers must bring Hawk to justice. At trial, however, Hawk does not recognize the court and denies its jurisdiction over him. When verdict and sentence are imminent and Rogers attempts to intervene, referring to Hawk as a man, Hawk solemnly reminds them all, "I am not a man." He then distinguishes law and justice, leading to a consensus to free him if the senior officer will guarantee his conduct. Characteristically (for himself and his lineage), Hawk refuses to submit to any conditions on his freedom. The impasse is resolved when he and Rogers agree to work together, since both are "aliens," Hawk from his own people and Rogers from his own time.

NOTES

1. 1952, George Marshall. Based on a novel by L. L. Foreman. See Rovin, *The Films of Charlton Heston* (1977).

2. Quoted in Rovin, 13.

3. 1954, Robert Aldrich. Based on the novel *Broncho Apache* by Paul I. Wellman.

4. 1983, Ted Kotcheff. Based on the novel by David Morrell. The sequel is *Rambo: First Blood Part II,* 1985, George P. Cosmatos.

5. 1968, Franklin J. Schaffner. Based on *Monkey Planet* by Pierre Boule; see Rovin.

6. George, *Julie of the Wolves,* 1972.

7. Rand, *The Fountainhead,* 202.

8. Ibid., 15–17, 35.

9. Ibid., 26, 32–34, 36, 89.

10. Ibid., 90, 92–93, 136.

11. Ibid., 308.

12. Llywelyn, *Finn McCool* (1994), 15, 17.

13. Ibid., 22, 38, 79.

14. Ibid., 127.

15. Ibid., 141, 158.

16. Ibid., 187, 189, 207, 215.

17. Ibid., 220.

18. Stewart, *Earth Abides* (1949).

19. Guild America Books (*Seventh Son,* 1987; *Red Prophet,* 1988; *Prentice Alvin,* 1989).

20.   *Red Prophet*, 248–249.

21.   Ibid., 249.

22.   Ibid., 272.

23.   Card, 300–302.

24.   *The Swords Trilogy*, 1971, New York: Berkeley Books 1982 printing (*The Knight of the Swords, The Queen of the Swords, The King of the Swords*).

25.   *The Knight of the Swords*.

26.   Ibid., 45.

27.   Ibid., 103.

28.   See, e.g., *The Queen of the Swords*, 284.

29.   Ibid., 251.

30.   *The King of the Swords*, 284, 298.

31.   Dick, *Do Androids Dream of Electric Sheep?* (1968). This novel is the source of the latter-day film noir, *Bladerunner*, which in its original theatrical version (neither the version available on videotape nor the more recent "director's cut") is a highly affecting version of the expanded Wild Man theme.

32.   *Daughters of the Dolphin* (1968); *Destiny of the Dolphins* (1969), New York: Ballantine Books.

33.   *The Book of Skaith: The Adventures of Eric John Stark*, 1974, 1976, New York: Nelson Doubleday (Book Club edition) (*The Ginger Star, The Hounds of Skaith, The Reavers of Skaith*); *The Ginger Star*.

34.   Ibid., 54, 47.

35.   *The Hounds of Skaith*, 255.

36.   Brackett, *The Secret of Sinharat* (1964).

37.   Ibid., 6–7.

38.   *The Ginger Star*, 110.

39.   Ibid., 54.

40.   Brackett, *People of the Talisman* (1964).

41.   *The Secret of Sinharat*, 12.

42.   *The Ginger Star*, 28.

43.   Ibid., 60. Also see *People of the Talisman*, 88.

44.   *People of the Talisman*, 21.

45.   *The Ginger Star*, 53.

46.   Yarbro, *Hotel Transylvania* (1978).

47.   Yarbro, *Blood Games* (1979).

48.   *Hotel Transylvania*, 12.

49.   Ibid., 114, 70.

50.   Ibid., 107.

51.   Grossman, *Saturday Morning TV* (1981).

52.   1976, William Hale.

53.   Episode first aired on January 12, 1983.

54.   Brooks and Marsh, *The Complete Directory to Prime Time Network TV Shows 1946 — Present* (1979).

55.   Brooks and Marsh, *The Complete Directory*, 334.

56. Nimoy, *I Am Not Spock* (1976); Nimoy, *I Am Spock* (1995).
57. Whitfield and Roddenbury, *The Making of Star Trek* (1968), 125.
58. Ibid.
59. Lichtenberg, Marshak, and Winston, *Star Trek Lives!* (1975), 220.
60. Whitfield and Roddenbury, *The Making of Star Trek*, 227.
61. Lichtenberg, Marshak, and Winston, *Star Trek Lives!* 49–50.
62. Woodley, *Man from Atlantis* (1977).
63. Fontana, *The Questor Tapes* (1974).
64. Hudis, *Time of the Hawk*, undated reediting of the original episodes.

V

STEWARDSHIP: RECONNECTION WITH THE NATURAL WORLD

16

The Human Niche and Defining a Personal Environmental Ethic

PLACE

When we were hunters in pursuit of survival our lives were more than physical; they were holistic, including something beyond the material realm. While we cannot reclaim that version of human nature, we can seek restoration of an "interpenetration of man and nature." To do so, we may have to explore beyond the boundaries of our relentlessly quantitative natural sciences, beginning with a "fundamental human attitude toward the landscape" beyond science into the mystery of our relationship with place. If we trouble to perceive the natural world as does the hunter, primal and perhaps modern, we may be moved by an awe shared throughout our species and its history.[1]

In a way we would be returning home. The irony is that we establish wildlife refuges for all creatures but ourselves. Even though we count loss of habitat as the greatest disaster befalling a species short of extinction itself, we allow ourselves to be lost in "cities, ranches, highways, noise, and other dissonance" as though we need no natural habitat. Yet we know that to survive animals "must at least from time to time have a home place, a place where they feel both protected and free." We are inclined to relegate habitat to the vacation spot, where we can only visit, but rarely remain, despite finding ourselves reinvigorated by at least the prospect of the process. We may be living a metaphor for primal cultures, where initiates are swept to a different world and return to the "outer world" in a bittersweet mood, glad, refreshed, but wistful.[2]

As for environmental crises; they begin and end within each of us. "We alter the equation when we alter our relationship to the planet. We get connected when we act connected." To be connected it is essential to see the cosmos in a "new" perspective, that of story, storyteller, and listener. And then we need a new aesthetic that "actually succeeds in connecting humans to nature," an aesthetic that is "compelling, engaging, incredibly unifying, and gentle all at the same time." Such an aesthetic will merge science with art, in expressions and connections.[3]

We are requisite dwellers in ecosystems and interdependent with "the complex structure of the land and its smooth functioning." We know that a change in one component demands adjustment in the others. We are aware that our tools have enabled us "to make changes of unprecedented violence, rapidity, and scope." Yet we further know that the less violent our changes, "the greater the probability of successful readjustment."[4] We are nothing if not inconsistent through our ambivalent relationships with the natural world.

The mundane expressions of our search for relationship with landscape — place — can be found in the planting of an English lawn, the sowing of a wildflower garden, the establishment of a woodland path, the creation of a lily pond. In these simple avocations we may well be seeking a meaning so profound it is hidden to us. Gardeners (and farmers) may be eminently practical, but we all are searching out "connections and even visions . . . not alien to our sensibilities."[5]

Today our landscapes are almost certainly artifacts. And they reveal our "culture and traditions as directly" as any novel, newspaper, or codified law. "Even the 'natural' or unaltered, parts of the planet are increasingly the result of human choice. Wilderness remains because we allow it to exist — in national parks and preservation systems." "[N]ature is not only a document revealing past thought and action but also a slate upon which the present outlines the kind of world it bequeaths to the future."[6]

And yet we almost inevitably claim individually to *want* something very different. The city-dweller may feel isolated in the metropolis by both vocational need and a mistaken sense of real detachment from the natural world, but the latter is impossible. Not only does the larger world physically sustain the most entrenched urbanite but it also nourishes the soul.[7] Even New York City is enhanced by Central Park. When it was founded, real wilderness was actually near to hand. The park was intended to be a "romantic improvement on the wild, a carefully fashioned landscape where city dwellers could come and enjoy the illusion of wilderness without any of its inconveniences or dangers." Now it serves in a new role, with pristine wilderness "a rare commodity." But there is hope here as Central Park becomes home to wild birds and their watchers.[8]

But even the rural landscape of the early twentieth century bore the brunt of human depredation. Laura Ingalls Wilder, in her column, wrote of a depleted farm with its poor soil, erosion, and land ruined by the altered course of a creek. On behalf of all of us, she admonished, "We are the heirs of the ages; but the estate is entailed, as large estates frequently are, so that while we inherit the earth, . . . we have only the use of it while we live and must pass it on to those who come after us. We hold the property in trust and have no right to injure it or to lessen its value. To do so is dishonest, stealing from our heirs their inheritance."[9] Thus, this popular author repeats the mutual thought of many ecologists, environmentalists, and the author of Genesis. And she goes on to the timeless assertion: "It should be a matter of pride to keep . . . that little bit of the earth's surface for which we are responsible in good condition, passing it on to our successor better than we found it."[10]

We miss at least the connection to the earth, if not the unavoidable physical realities. Some go beyond the cultivated pastoral image to return all the way to the wild, at least in mind and spirit.[11]

ARCHETYPES AND SURROGATES

One way or another most of us seek the pastoral, the atavistic, or some expression of our seemingly abandoned niche through archetypes and surrogates. Anthropologist Tobias Schneebaum puts the search in words:

New Guinea had stood out as my final hiding place. There was nowhere left for me to explore myself, to look for the wild man. . . . The wild man allowed no frivolities; he was ugly and in my concept of him was biologically, anthropologically, evolutionarily, paleontologically, Primitive Man. His surface was as violent as his interior was gentle, and the very looks of him were fierce. . . . I could displace everything, rearrange my lives, replace my past with that of the wild man, instill his presence into my void, and stuff his integrity into my despair.[12]

As Schneebaum observed primal peoples fishing he describes "feeling back to a time when man's relationship to his surroundings was vital to his being, when he absorbed his life and blood from the land and the spirits with whom he lived."[13] Thus it seems we are missing from our niche three nourishing sources: the connection to it through our very blood and life, the spiritual realm, the ubiquitous vitality itself. We want to be like our archetypes, to achieve their accomplishments and follow their trails. In terms of the Feminine, the Anima within each of us, man as well as woman, "it is the way of retaining and developing" our souls. The "Wild Woman is the one who dares, and who creates and who destroys. She is the primitive and inventing soul that makes all creative acts and arts possible. She creates a forest around us and we begin to deal with life from that fresh and original perspective."[14]

The alternative to embracing our collective unconscious with its wealth of archetypes is to fall prey to "low self-esteem, addictive behavior, and neurosis, so the culture that disregards its group dreams falls prey to an alienation that, at least at the present moment, manifests as gluttony, selfishness, and possibly self-destructiveness."[15]

Some of us find an outlet in our dreams, where even the most pedestrian of us "leave earth behind and go flying." What are we seeking when "[t]he tears of dreams can be real enough to wet the pillow and the passions of them fierce enough to make the flesh burn"? There are certainly times when "we dream our way to a truth or an insight so overwhelming that it startles us awake and haunts us for years to come." While everyone from scholars to fortunetellers to outright phonies proclaim their meaning — and others dismiss them as meaningless — dreams almost certainly reveal that we are "in constant touch with a world that is as real to us while we are in it, and has as much to do with who we are, . . . as the world of waking reality." The other revelation is that "our lives are a great deal richer, deeper, more intricately interrelated, more mysterious, and less limited by time and space than we commonly suppose." It may well be the material world is not the exclusive reality.[16]

It is time we heeded such dreams, sleeping or waking. It is time to expand our notion of ecology and reclaim our niche.

HUMAN ECOLOGY

In keeping with the point of view of ecological scientists, human ecology refers to how the population uses the environment through the apparatus of cultures. We are beginning to recognize and appreciate that at least some Amerindian peoples were (and some remain) highly accomplished practical botanists and zoologists engaged in "creative stewardship." Their cultures exhibit an awareness of the fragility of bonds connecting the pieces of nature and the need to avoid injury to those bonds, if only to avoid harm to themselves.[17] While we may have no desire to resume a primal way of life, we are yearning for something at least approximating that human presence in the species' niche. As a species, we seemingly have progressed beyond Neanderthals and their use of tools and artifacts into a form of abstract reasoning, including a sense of aesthetics,[18] which should serve our need for reconnection.

We like to deem ourselves remarkable in possessing a brain and social life comparable to the dolphin's, along with the faculty of speech and use of tools. Our cultures can be deemed a new level of emergent organization tracing back to pure mathematics, physics, chemistry, and biology, incorporating each along the way. We have the innate skills that can aid us in reconnecting with our ecological niche and using our techological niche to do so. What sets us apart as uniquely unique among species is our ability to change our niche at will. Our physiques, temperaments, subtle desires, and behavior patterns still reflect our niche as wandering ice-age foragers. There would appear to be no reason we are unable to narrow our present ubiquitous niche a bit in search of reconnection. A lesson of ecology to be taken to heart is that all poverty is attributable to unrestrained population growth,[19] especially where that growth transcends mere numbers.

Our species has been called the moral animal. Certain ethical standards may be innate, not merely learned. Certainly possession of an (inchoate) ethic is inherent in us.[20] With the prospect of engagement in a personal environmental ethic we are beginning to put our unique features together in an effort to restore our relationship to the natural world.

We need a new awareness on any number of levels. After all, when we avail ourselves of the instruments and gadgets our technology provides, we seldom pause to "recite the social, political, cultural, or health-related consequences"; it might even be "impossible to function if one were constantly questioning a machine's effect in society at large" or beyond.[21] But we might occasionally pause for reflection. Contrary to what might once have been a cultural belief, "[t]he earth is not our enemy, or something we must outwit, but it is our source, our Mother, essentially good and to be trusted."[22] Indeed, to set neither goal nor time, but to take place and situations as they come is actually a profound release; we should allow natural patterns to filter into our consciousness[23] — witness both those dreams and those archetypal surrogates of our popular arts.

As a species we can be deemed a new power of the universe itself; with the power of consciousness giving the universe the ability to turn back and reflect upon itself. Fire is often deemed the source of our human-ness.[24] Before a fire "strange stirrings take place within" us, and "a light comes into [our] eyes which was not there before." An open flame draws us into an environment of adventure and romance. It is a contact with the past, a recapturing of "the lost wonder of . . . early years and some of the sense of mystery of [our] forebears. . . . Having bridged the gap," we are quick to discover something lost, "a sense of belonging to the earth" as well as to our kind. Fire, in short, remains "a primal and psychological necessity."[25]

TRANSCENDING HUMAN ECOLOGY

Lynn Margulis and Dorion Sagan advise that a species grows until it challenges the carrying capacity of its ecosystem, whereupon it either becomes extinct or it transcends itself.[26] Like other life, our species has undergone evolution. And evolution's "first grand lesson" is "the unity of life." It is no accident that early Christianity among other religions grasped the principle of the "brotherhood of man," a truth confirmed and expanded by evolution: "all living things are brothers in the very real, material sense that all have arisen from one source and been developed within the divergent intricacies of one process." We are a part of nature and "kin to all life."[27] But we are uniquely unique. "We must discover what it means to search for an equilibrium between the polarities of nature and God."[28] The questions of whether we are an image of God or merely some machine or computer are far from merely trivial or purely academic.[29] Indeed, the lesson to be learned may be that there is no polarity after all.

There is a new danger to heed: the millions of years of evolution of the human body and the thousands of years of evolution of human culture do not seem to work at the same rate. "Our bodies can't adapt fast enough to the different environments that society and technology are constantly creating."[30] This realization alone should move us to consider something beyond the culture of this time and the exclusively material realm. The collective unconscious and the human propensity for an ethic are far older than our culture. At the very least we should explore all our cultural threads, not just one here and another there. There is a wealth of variety in the colors to be woven together in a new tapestry.

It is incumbent upon us to develop a new law since we believe we are capable of circumventing evolutionary law and ecological laws of niche. We must learn and exercise restraint. Our uniqueness, after all, arose from "sun-driven protoplasm" and our genius as a species grew out of "an intimacy with the earth." Should we not go back to find both the intimacy and the genius?[31]

In the end, society's — a culture's — character "is the cumulative result of countless small actions, day in and day out, of millions of persons. Who we are, as a society, is the synergistic accumulation of who we are as individuals."[32] If we can reconnect as individuals, at length our culture will necessarily follow our lead.

We can anticipate improvement in cultural relationships to nature "when we elevate our poets, musicians, shamans, and philosophers to the critical position occupied by our scientists and politicians" in "defining and explaining nature for the rest of us." Today scientific ecology is already joined by a spiritual ecology that may serve to carry us across generations of past and future. This ecology "asks us to support the ghost dancers as well as the nuclear priesthood. It invites us to venture out into nature and attune ourselves with the migrating songbirds, who provide some of the loveliest accompaniment to the generational dance."[33]

A new ethic will not emerge on its own, subject to the kind of decision that selects a new coat. It must emerge out of existing attitudes. Similarly, a morality or religion is not to be conjured but must grow from existing attitudes, an extension or a development. And ethics, morality, or religion lacks any point if it fails to govern conduct.[34] If our new approach to the natural world is to have so profound an influence, four essentials are in order: a balancing among human, social, and environmental needs; improved methods for assessing environmental conditions; the ability to distinguish between improvement and perfection in establishing realistic goals; and "self-discipline, commitment, and a willingness to cooperate." Instead of our present adversarial pathways to resolve the local and the global environmental matter, we need a "workable, consensual approach, involving all parties with a stake in the outcome." To that end, given the preferred subservience of government to the people, we need to accept the lesson of self-discipline, as must any responsible sovereign.[35]

The earth will not be saved through dictating laws in which we may neither believe nor care. "We must change the frequency of the human mind."[36] Even an attorney can concur: Despite the complexity of contemporary environmental laws, they are an "incomplete response to the moral and scientific lessons of ecology." The most egregious problems are the preclusion of holistic solutions and the slighting of aesthetic values. A new ethic is essential to bridging the gap between knowledge and action.[37] The notion is hardly novel: "Intrinsic values underlie, and always have underlain, much of the law made by the legislature and the courts." Law, then, merely sets the minimum, not the optimum. It is up to us to apply higher standards. Here is the distinction between mere law and morality.[38]

What our environment demands of us may be the "most complete reversal of values that has taken place since the Neolithic period,"[39] in short, transcendence of our own nature. We should never give up hope; if we see "the beauty in nature," we are far more likely to work for its preservation before it is too late.[40] A new ethic must be taught in kindergarten and thereafter.[41] Ecological literacy already begins in childhood, when "the sheer delight of being alive in a beautiful, mysterious, bountiful world" drives our sense of wonder.[42]

The biophilia hypothesis proferred by Stephen R. Kellert and Edward O. Wilson asserts our spiritual and aesthetic dependence on nature, going beyond our innate ecological dependence. The hypothesis would carry us "beyond the poetic and philosophical articulation of nature's capacity to inspire and morally inform" to a scientifically acknowledged need, "fired in the crucible of evolutionary development, for deep and intimate association with the natural environment, particularly its living biota."[43] We already have the inspiration for nothing less than a new world order founded in science and ethics, a new science "found in all religious and spiritual traditions." It can serve us in building "on our emerging awareness of our . . . species as a conscious part of the earth, . . . a living planet . . . , the mysterious self-organizing Earth Mother, nurturer of us — and of all life."[44] Many of us are already finding some outlet for a "pervasive spirit . . . distinctly modern and Western, a braiding together of folklore, environmental research, and the news of the day. . . . [A]ll are asked to 'remember our bio-ecological history, as our species and its forebears evolved.'"[45] Now that spirit and our eagerness to respond must be tapped. Individually and as a culture we must join the scientific world of reality with "our capacities for that intimate communion with the natural world that has inspired the human venture over the centuries," one with "poets, musicians, artists and spiritual personalities."[46]

The closing chapters seek to gather an array of our intellectual assets to bring to bear on our political system as well as ourselves in the name of stewardship as a renewed way of life. Our realm is the cosmos and its interpreters, whether scientist, artist, clergyman, or statesman. That realm is of the spirit as well as of subatomic particles and physical energy. It is in a very real sense a part of our popular culture.

NOTES

1. Shepard, *Man in the Landscape* (1967, 1991), 208, 211, 213.
2. Estés, *Women Who Run with the Wolves* (1992), 266–267, 290.
3. Nollman, *Spiritual Ecology* (1990), 196, 198, 201.
4. Leopold, "Land Ethic," in Leopold (1966), 253–254, 257.

5. Fenyvesi, "The Power of Landscapes," *Washington Post* (Valley News), June 21, 1995, C1.

6. Nash, *American Environmentalism* (1990), 1, 2.

7. Nollman, *Spiritual Ecology*, 204–205.

8. Winn, "Central Park and Its Bird-Watchers Go Wild," *The Wall Street Journal*, September 8, 1994, A16.

9. "Heirs of the Ages," November 1923, reprinted in Hines (1991), 47.

10. Ibid.

11. See, e.g., Snyder, *The Practice of the Wild* (1990), 180.

12. Schneebaum, *Wild Man* (1979), 196–197.

13. Ibid., 229.

14. Estés, *Women Who Run with the Wolves*, 113.

15. Nollman, *Spiritual Ecology*, 183.

16. Connor, *Listening to Your Life* (1992), 213–214.

17. Heizer and Elsasser, *The Natural World of the California Indians* (1980), 57, 59.

18. Gould, *Wonderful Life* (1989), 320.

19. Colinvaux, *Why Big Fierce Animals Are Rare* (1978, 1979), 222.

20. Simpson, *The Meaning of Evolution* (1949, 1951), 144.

21. Mander, *In the Absence of the Sacred* (1991), 33.

22. *Earth Wisdom,* 53.

23. Mosher, *Songs of the North* (1987), 135–136.

24. See, e.g., Swimme and Berry, *The Universe Story* (1992), 143, 149.

25. Mosher, *Songs of the North*, 217, 218, 219.

26. Margulis and Sagan, *Microcosmos* (1986), 243.

27. Simpson, *The Meaning of Evolution*, 13–14, 136.

28. Shepard, *Man in the Landscape*, 206.

29. Orr, *Ecological Literacy* (1992), 93–94.

30. Zeveloff et al., *Wilderness Tapestry* (1992), 11.

31. Lopez, *Arctic Dreams* (1986), 39–40.

32. Elgin, *Voluntary Simplicity* (1981), 22–23.

33. Nollman, *Spiritual Ecology*, 14, 20.

34. Passmore, *Man's Responsibility for Nature* (1974), 56, 111.

35. Brunner et al., *Corporations and the Environment* (1981), 4, 9, 31.

36. Mien, quoted in McFadden (1991), 234.

37. Tarlock, *"Earth and Other Ethics*: The Institutional Issues" (1988), 44, 76. The reference in the title is to the book by Christopher Stone, with the subtitle, *The Case for Moral Pluralism.*

38. Kempner, "The Environment: God's Creation, Human Regulation," written notes for Feen Lecture Ohavi Zedak Synagogue in Burlington, Vermont, September 8, 1993, 15, 18.

39. Berry, *The Dream of the Earth* (1988), 159.

40. Wyland, in Allen, *Biosphere II* (1991), cover.

41. Douglas, *My Wilderness* (1961), 33.

42.   Orr, *Ecological Literacy*, 86.

43.   Kellert and Wilson, *The Biophilia Hypothesis* (1993), 20–21.

44.   Henderson, in Bruner, Miller, and Stockholm (1981), 51, 53. See also Weiskel, "The Need for Miracles in the Age of Science" (1990), unpaginated.

45.   Roszak, *The Voice of the Earth* (1992), 245.

46.   Swimme and Berry, *The Universe Story*, 2.

17

Cultural Sources for a Restored Relationship

WEAVING THE CULTURAL THREADS IN A NEW TAPESTRY

A culture's story of its universe and the human role in it is at the very least a primary source of intelligibility and value. The story instills appreciation for life's meaning and is a source of psychic energy to deal with the crises in the lives of both the culture and its individual members. Thomas Berry and other visionaries are calling our culture to a new historical vision, one of an intimate earth community of human, biological, and geological components. Notably, the legitimacy of the call is suggested by natural science itself, which is leading us out of a specific lack of knowledge coupled with insufficient intimacy "in the great family of the earth,"[1] both ironically attributable in part to science itself, once nearly exclusively reductionist and positivist.

The time has come when we not only can but must listen to the stories extending our own existence and seek fulfillment in our unbreakable connection with the world. Most of us have "forgotten our primordial capacity for language at the elementary level of song and dance, wherein we share our existence with the animals and with all the natural phenomena." Perhaps our fascination with First Nation cultures is deeper than some suspect in representing connections "we still carry deep within ourselves, beyond all the suppressions and even the antagonism imposed by our cultural traditions."[2]

The acknowledged visionaries are not alone in the call for a new ethic. Visions, however constructed, are the foundations upon which theories depend. According to free market environmentalists human nature is such that "good resource stewardship depends on how well social institutions harness self-interest through individual incentives." This movement notes the gap in knowledge between experts (or scholars) and most individuals who would take personal action. As a result no prescription for solution is offered. Instead process is suggested, specifically "linking wealth to good stewardship through private ownership" by which the market process "generates many individual experiments" with the successful experiments later copied.[3]

If our diverse perspectives can reach this much concurrence, the time of the underlying foundation and emergent ethic has come. The material realm is not the sole place of that ethic. As a major element of our culture, the Judeo-Christian tradition has a legitimate role in responding to the problems. Even so pragmatic a source as *The Wall Street Journal* has discovered a rabbi who combines faith with energy conservation and teaches that environmental protection is deep in Jewish law and tradition.[4]

Another point of departure is the Gaia Hypothesis. The modern concept of Gaia was coined by novelist William Golding who shared it with James Lovelock. However conceived and whether either necessary or sufficient to serve as foundation to environmental protection, the Gaia Hypothesis is yet another expression of our need for reconnection and one that reveals yet again that science and religion and even magic are themselves more intricately interconnected than our culture at large may be prepared to accept.[5]

COSMOLOGY: THE MATERIAL REALM

At this juncture it is helpful to review something of the history of the "new" cosmology as a crucial component of contemporary science in its willingness to accept complexity and holistic views of the material realm.[6]

In the 1920s Hubble's identification of the Andromeda nebula as a galaxy provided a sense of size to infinity. By 1930 red shifting in galaxies represented the "first solid evidence of cosmic expansion." In the 1950s Hoyle coined the phrase "Big Bang" in derision. In the years since, scientists have come to believe they have found a variety of evidence for the reality of the Big Bang and the consequent expanding universe. Because of the intricate and intimate interconnection from pure mathematics to physics to chemistry to biology to ecology and beyond, cosmic history is in effect "the history of everything there is," and everything, including the subnuclear particles originating in the Big Bang, was created, has matured, and evolves. The interconnections are truly phenomenal, starting with the infinitely immense and the infinitely minute, quasars and the Big Bang, quantum mechanics and Einstein's relativity.

Scientists themselves are growing in *overt* appreciation of the significance of nature's structural hierarchy. They no longer explain away "holistic concepts like life, organization and mind" as merely "atoms or quarks or unified forces or whatever. *However important it may be to understand the fundamental simplicity at the heart of all natural phenomena, it cannot be the whole story. The complexity is just as important.*"[7] Part of the importance is in the depth of organization and self-regulation in systems. The potential for both, too, may have been present at the Big Bang. The two may be so deeply embedded in nature that we may find them "functioning among the seemingly most chaotic phenomena. Where reductionism and materialism could find continuity between the human and the natural only at the physical level, we can now imagine a higher connection at the level of mentality."[8] In short, science is not only providing "some of our most powerful poetic references and metaphoric expressions," but scientists are now able to express their awareness of a "magic quality of the earth and of the universe entire" among their colleagues.[9]

That only the overt expression of such perceptions is novel is suggested in the association of beauty with science, at least among scientists. Theories are often described as beautiful. Since our roots among the Greek philosophers, beauty has been perceived in symmetry. The scientific standard of beauty has altered from time to time, but recently Einstein's equations, among others, have been deemed especially beautiful.[10] The beauty of the cosmos is both emotional and intellectual. For many it is spiritual as well.

Even absent poetry, nature's delicate balances and seeming coincidences continue to maintain the systems making up the cosmos and our more immediate environments. The most casual study of either biology or ecology will confirm their ongoing presence. For those left unmoved by poetry, there is adventure. And that is a partner of science as well, though both are dangerous: "Science and adventure are our entry into the future, our way of telling succeeding generations that we are not afraid to open the door onto their lives. The risks, to be sure, are formidable. But the rewards are great, too."[11]

The lesson here is not to reject poetry, adventure, or science, but to make all of them an integral part of our own relationship with the world around us—and to learn to balance them as we proceed. It is worth remembering that an organism, as only one example, is not a machine that can be disassembled and restored to functional wholeness. Take an organism "apart and something special happens. It dies. Interrupt its vital processes for any length of time by pulling it to pieces and it loses something that must have been there hidden away in the relationship between the parts and which cannot be restored."[12]

Perhaps the same can be said of the cosmos as a realm of the spirit as well as of the material. Perhaps the time has come to put the two back together before the cosmos, or our small piece of it, expires. Seemingly, we are on many brinks today. One of them is "of a rapprochement between science and religion, or at least between science and certain religious categories of thought that have long been exiled from our secular culture."[13] After all, "[s]cientific thought, being only one of its functions, can never exhaust all the possibilities of life."[14] It is in our nature to be of and yet to transcend biology. We should not submit to being only half ourselves.

COSMOGONY: THE METAPHYSICAL REALM

It can come as no surprise that many theologians view the Big Bang as the key to reconciling cosmology with theology. An array of players—secular scientists, cosmologists, theologians, and creationist scientists and theologians—can accept creation as a one-time event in real time. For at least some of these players creation hardly stopped with the Big Bang but is a "continuous process, allowing for a highly orderly universe to emerge from a cosmic void." Significantly, the Big Bang and Genesis "do not offer different explanations for the same phenomenon, but answer different questions." The one answers how, the other who or at least why.[15]

It is a sign of the times that an organization such as the American Scientific Affiliation provides a scholarly outlet for individuals who are both scientists and Christians of a variety of denominations. The articles cited here are only a sample of the relevant writings to be found in ASA's journal. It is also a sign of the times that similar expression is to be found not only in *The Wall Street Journal* but also *Time* and a plethora of relatively popular books, some of which have been cited throughout this volume. *Time* concedes the possibility of religious interpretation for "some of the epic narratives of contemporary science" and notes a hallmark of our science in the amount of "fertile ground it has provided for bona fide theological speculation."[16] The article wonders about (if not over) the seeming calibration of the universe for life in the array of prerequisite conditions met. The article takes into account chaos theory and the insistent emergence of structure from chaos.

A sidebar article[17] notes the acknowledgment of Galileo's Christianity by Pope John Paul II, who refers to Galileo's timely admonition "that Scriptures cannot err but are often misunderstood." This sidebar summarizes the history of mutual regard between scientists and theologians, while the principal article asserts, "one of the great scientific minds of our era believes that the ultimate questions remain unanswered, that science may be unable to answer them, and yet that science does help us mull them over, by illuminating the epic trajectory of cosmic and biological evolution."[18]

In the context of chaos theory, a fellow of the American Enterprise Institute is among those who perceive a "yawning philosophical vacuum in American life, with people urgently trying to fill it in myriad ways."[19] A imminent "great transition" is seen behind a "host" of seemingly unrelated social phenomena, including increasing membership in evangelical religions, New Age activities, alternative medicine, an explosion of street gangs, the feminist movement, and the environmental movement itself. To this observer, each is representative of "a hunger for something beyond material progress." And there are risks along with the promise so many see in this opening of our minds: inward looking can lead to xenophobia, mysticism can devolve to fanaticism, stability can lead to authoritarian government.[20] "But there may be further hope in that — The world is watching: one cannot walk through a meadow or forest without a ripple of report spreading out from one's passage. The thrush darts back, the jay squalls, a beetle scuttles under the grasses, and the signal is passed along. Every creature knows when a hawk is cruising or a human strolling." Maybe they can help keep us on the right track—if we heed them. Notably, animals have many a theological as well as ecological niche.[21]

In fact, this separation between theology and science, including their respective roles in environmental affairs, is a curiosity of our historical era. Three classic queries of philosophy and religion are: What is the character of undisturbed nature? How does nature influence people? How do people influence nature? The question which may be uniquely ours to ponder is—How should they? We have yielded those questions to science and scholars and diminished the deeper imports of the questions in the process.[22] It is time to reclaim the questions along with the meanings deeply personal to us.

To blame some combination of industry and government for environmental problems is facile. To expect reconciliation and restoration to spring from them is unrealistic. A Jewish religious leader and scholar points to a "bad way of life" rather than bad politics and urges attacking the problems "from the bottom up, at the human level, . . . the challenge of religion." But the question is whether religion can imbue individuals "with a spiritual renewal that will ennoble [our] worldview, enlarge [our] inner life, and temper [our] wants." "In Judaism, freedom is a blessing that must be harnessed, structured, and channeled to become beneficent." Judaism provides a "sacred canopy of legal regulations, theological notions, and intellectual values on a bedrock of a transcendental God concept, the outcome of which is a decidedly modest sense of man's place and purpose in the universe." We are in effect God's tenants. The planet is His and we owe Him a duty of care.[23]

From the contemporary Christian perspective a corresponding ethic is suggested: Abuse of creation is wrong because it expresses willful rebellion against God. Creation is to be valued because of God's value of it. This emerging consensus among multiple spiritual perspectives demands that we preserve and cherish the earth. But radical changes in both individual behavior and public policy are prerequisite.[24]

Pope John Paul II asserts,

Creation was given and entrusted to humankind as a duty, . . . *the foundation of a creative existence in the world.* The person who believes in the essential goodness of all creation is capable of discovering all the secrets of creation, in order to perfect continually the work assigned to him by God. . . . [T]here is a great challenge to perfect creation—be it oneself, be it the world.

Creation is permeated with a redemptive sanctification, even a divinization.[25]

The Pope reminds us that Emmanuel means "God-with-Us" and expressly refutes "the completely false image of God which the Enlightenment accepted uncritically," a God who is *only* transcendent. According to the Pope, "God is not someone who remains only outside the world, content to be in Himself all-knowing and omnipotent. *His wisdom and omnipotence are placed, by free choice, at the service of creation.*" Others may refer to an immanent God, but, even more important in this context, "[f]or Christians, the world is God's creation, redeemed by Christ. It is in the world that" we meet God,[26] whatever form the deity takes for us.

NEW COSMOLOGY: INTEGRATION
FOR RECONCILIATION

Jungian Psychology

According to Jung, humankind of all times and places has inevitably established religious expressions. Our "psyche from time immemorial has been shot through with religious feelings and ideas." To dismiss this part of ourselves is to be blind, "and whoever chooses to explain . . . or to enlighten it away, has no sense of reality." Jung urges us not to turn back "the wheel of history" but to continue our "advance towards a spiritual life, which began with the primitive rites of initiation" and which "must not be denied." To him the "human psyche may not be parcelled out" as science persists in pursuing, because intellect is only one of the functions of consciousness. Despite thousands of years of rites teaching spiritual rebirth, we continually forget "the meaning of divine procreation." Spirit may be easily driven "out the door," but without it life "loses its savor." To Jung the spirit is the living body "seen from within" and the body is the "outer manifestation of the living spirit": the two are one. Nevertheless, in our world there is an "enormous tension" between outer and inner life and between objective and subjective reality.[27]

The Rationale for Integration

If Jung tells us body and spirit are one, Berry is telling us that the destinies of the universe and humanity are a unity. Berry rediscovers a common basis for teaching in science and technology and in religion and humanities and a mutual appreciation between these diverse lines of inquiry. Here he finds "our greatest promise for the future," but it is also "the great task of the educator" who must first comprehend and then communicate "this vision to future generations of students." It is only now that we can see through the implications of the results of scientific inquiry that we live in "a cosmogenesis best presented in narrative; scientific in its data, mythic in its form." Amazingly, the more precisely we quantify at the scale of relativity and at the scale of quantum mechanics, the less precise and more chaotic the order portrayed by science actually becomes. It is reductionism itself which has "brought the human mind into a new depth of understanding."[28]

It is ironic that the realization of contemporary physics that "everything has a quantum interconnectedness" is at least akin to what Amerindians have been holding all along. Time means something entirely new to us.

And space is never empty. Some hope the awareness of these revised precepts will permeate our culture. That awareness may already be "one reason why people write and . . . paint and . . . share things with each other, dance, try to see these connections."[29] In this wondrous new context, a new question, of course, arises: "What are the long-range consequences human activity will have on cosmic evolution?"[30]

Contemporary Christian Stewardship

The Judeo-Christian tradition has been held accountable for human depradation on the planet. A reasonable query is whether a Christian concept of stewardship is among the inspirations for a sound environmental ethic. Among those who challenge both Christians and the institution of Christianity is Carl Sagan, who, in turn, is criticized for "rarely, if ever" distinguishing "the intentions of Christianity's founder" and the way we have sometimes worked things out with little apparent regard for those intentions. This failure, Sagan's critic argues, is akin to arguing the invalidity of the scientific method because some scientists use the resultant premises to develop weapons of mass destruction.[31]

Perhaps it is time to afford our own tradition a fresh look. Neither interpreting nor reinterpreting biblical messages is a novel practice. One such endeavor uses Sagan's own words to make the point of a connection between conclusions of science and theology: "Deep, inpenetrable darkness was everywhere, hydrogen atoms in the void." To this commentator "scientific evidence and holy scripture must harmonize . . . because they are evidence of the same creation by the same Creator." It would be impossible for scientists to produce "facts we do not want to hear" because God's universe and His "holy word would not contradict" each other. "It is humankind who is confused."[32]

One who actively practices ecological Christianity asserts life to be an open-ended journey in which we are seeking neither certainty nor an end to ambiguity but instead "depth, breadth, and meaning." Those Christians in what Jay McDaniel identifies as a kind of independent "third-phase Christianity" are in an ongoing process of spiritual change in search of enrichment and sustenance—and the living God—without the binding of doctrinal compartmentalization. For him this is the spiritual place of ecological Christianity.[33] This may also be the place for the independent spirit who would not align with a formal religious institution.

Still other Christian commentators call for an ecological theology introducing the elements of immanence, the relationship of humankind to creation at large, and the role of the church itself in environmental affairs.

These commentators challenge Christians to fulfill the role of steward-ecologist. Their article reports on a survey following the Lynn White challenge to Christians (not Christianity) in which theologians asserted that God is immanent and cares for all creatures, citing the examples of Matthew 6:26 and Psalm 104:21–31 and Romans 1:20 for the proposition that nature is "animated, ruled and sustained by the 'breath of God.'"[34]

The covenant extended to the children of Abraham is *among* God, humans, and nature (Genesis 9:8–17), and nature is wounded by the Fall (Genesis 3:17; Romans 8:20). The incarnation of Christ "evidences the goodness of creation" (John 1:3–4; Colossians 1:17). All of creation, not merely Christians or humanity, awaits redemption (John 8:16; Colossians 1:20). Isaiah certainly suggests an ultimate reconciliation among humanity and both animal and plant life (Isaiah 11:6–9; see also Ezekial 34:25–27). Thus, there is every indication (upon a close study of the Bible akin to that scientists or lawyers afford their respective texts) that "humans are ontologically one with nature" and that "the relationship between humans and . . . creation must proceed from that fact." To these commentators to be a Christian is to accept stewardship of creation (from within), which duties demand one to be at some level an ecologist.[35]

In anticipation of an array of new integrations in the name of restoring our relationships to the cosmos, theology, if nothing more, is an important source of inspiration. Deliverance will come from the many directions of our quests, including those for the aesthetics as well as the scientific. While theology cannot prescribe answers to questions raised elsewhere, it can avail itself of those answers in pursuit of its own quests. Once the Queen of the Sciences, theology's crucial role now "lies in its commitment to seek the deepest possible level of understanding."[36]

Not only Christians or those of the Judeo-Christian tradition long for a new, inclusive peace with Earth and animals. The contemporary Christian or at least institutional Christianity is belatedly discovering that the global village exists and extends beyond the human abode. We are fellow pilgrims who are claiming the earth as home—and *shalom*. Shalom is dynamic, both the absence of violence and the fullness of life. Shalom is relative; rattlesnakes will bite and earthquakes rattle. But the way has been paved by eons of geological and biological evolution. "To be able to partake of this shalom . . . is to live up to the idea that we are made in God's image. Even God, we believe, is a God of shalom. Our dream of shalom is our way of responding to a divine dream, that the will of God might be done 'on Earth as it is in heaven.'"[37]

The ecological theologian of whatever sect or inclination will criticize anthropocentrism, recognize the human psyche and body as part of the web of life and the evolving cosmos, impose on believers an ethical obligation to Earth and other forms of life, and recognize the connectedness with Earth and life as significant and essential to spirituality.

Jay McDaniel offers an analogy: All creatures are like fish swimming in an ocean called God. This ocean is not merely external watery substance, but itself a living substance with an inside. Both fish and ocean feel, perceive, and dream. The ocean, however, is not located in a particular body. It is everywhere at once. It feels everything each fish does. More, it seeks to reconcile each fish "into a deeper pattern, a single whole."[38]

POLITICS AS LOOM

Returning to the analogy of a tapestry, the loom on which the tapestry of a new environmental ethic, personal and for the nation, is to be woven is a prosaic one: the politico-legal system. But our system can be one of caring, informed of cause and effect by science and of the meaning of kinship by philosophy, because "facts about the universe are no substitute for a philosophy of caring for life on earth."[39] Ecology defines the economic framework provided by the biosphere, ethics suggests how that wealth should be shared, aesthetics and religion (or spirituality) provide values for setting upper limits on material consumption. Technology and industry make abundance available, but it must be "democratically shared before there is a chance for it to be individually squandered." To avoid inane proliferation of culture's neon telephones, the proper response is "to tap . . . deeper motivations, both aesthetic and religious."[40]

The most glaring weakness of reform proposals is their "omission of a concept of citizenship and participation in the process of change." Significantly, those who are capable of transforming leadership may well be those "rooted in place" but with loyalties extending to the planet. To open the door to these visionaries it will be necessary to "kindle the moral energy that will ignite the actions necessary to build" the anticipated future.[41]

The dream is one of daunting optimism. Hope may be found in the observation that the "notion that the public has a right to expect certain lands and natural areas to retain their natural characteristics is finding its way into American law." With a foundation in public trust doctrine, the idea is moving from the original navigable waters to increasingly distant resources, from other bodies of water to old growth forests to mountains and even to wildlife. There is some suggestion — in California case law and the social and political forces influencing at least one decision in that State, in scholarly commentary on public trust, and in unconventional sources for theory — that a public property right in ecological preservation may emerge.[42]

A caveat here lies in democracy's vocation for accommodating multiple or conflicting voices. The treasured system itself "would seem to rule out the possibility of creating anything coherent, much less lucidly harmonious. A further difficulty in reconciling democracy with the aesthetic is that the aesthetic implies excellence, a hierarchy of values, a focal point or several focal points of interest around which lesser ones are arranged."[43]

One prophetic example may be in the controversies engendered by the Endangered Species Act, described as possibly "the strongest official means for calling public attention to the deleterious effects of major environmental intervention." In practice this statute has turned the dispute into a claim of an individual, the species, against "powerful and hostile authorities, when actually what we want to preserve are places, landscapes, and fields of interaction between many species and natural contexts." This result is attributed to the failure of captivating the public imagination intuitively or through policy itself.[44]

And so it is with considerable trepidation that we turn our attention to the politico-legal system where all the philosophies and dreams of an environmental ethic come for expression on our behalf and practical application.

NOTES

1. Berry, *The Dream of Earth* (1988), xi, xiv.

2. Ibid., 2, 3.

3. Anderson and Leal, *Free Market Environmentalism* (1991), 4–5.

4. Aeppel, "For This Rabbi, Saving the Earth Just Comes with the Territory," *The Wall Street Journal*, June 1, 1994, B1. That this Rabbi and active members of the Christian tradition are hardly alone is evidenced in the nearly seven hundred-page volume, *This Sacred Earth: Religion, Nature, Environment* (Roger S. Gottlieb, ed.), 1996, New York: Routledge. Passages from Judeo-Christian writings are found throughout, along with the seminal indictment of the tradition, Lynn White, "The Historical Roots of Our Ecologic Crisis," reprinted from *Science* 155 (March 10, 1967), 184.

5. Sagan and Margulis, "Gaian Views," in Chapple (1994), 6. For more on the intriguing science of this pair, see their *What Is Life?* 1995, New York: Simon & Schuster.

6. The following chronology is from Roszak, *The Voice of the Earth* (1992), 112–113.

7. Davies, *God and the New Physics* (1983), 225 (emphasis added).

8. Roszak, *The Voice of the Earth*, 134–135.

9.   Berry, *The Dream of the Earth* (1988), 16.

10.   Snoke, "Toward a Unified View of Science and Theology" (1991), 172. *Perspectives* is the quarterly journal of the American Scientific Affiliation.

11.   Tucker, "An Ecological Cosmology," in Chapple (1982), 282.

12.   Roszak, *The Voice of the Earth*, 170.

13.   Ibid., 101.

14.   Jung, *Modern Man in Search of a Soul* (1933), 123.

15.   Philippidis, "Cosmic Controversy: The Big Bang and Genesis 1" (1995), 190, 192, 193.

16.   Wright, "Science, God and Man," *Time*, December 28, 1992, 38, 40.

17.   Ostling, "Galileo and Other Faithful Scientists," *Time*, 42.

18.   Wright, "Science, God and Man," 44.

19.   Farney, "Natural Questions: Chaos Theory Seeps into Ecology Debate, Stirring up a Tempest," *The Wall Street Journal*, July 11, 1994, A8.

20.   Ibid., A8.

21.   Snyder, *The Practice of the Wild* (1990), 19–20.

22.   Botkin, "Ecological Theory and Natural Resource Management," in Chapple (1994), 65–66.

23.   Schorsch, "Learning to Live with Less," in Rockefeller and Elder (1992), 30, 35.

24.   Van Dyke, "Ecology and the Christian Mind: Christians and the Environment in a New Decade" (1991), 174.

25.   Pope John Paul II, *Crossing the Threshold of Hope* (1995) 20–21, 22.

26.   Ibid., 62–63, 89.

27.   Jung, *Modern Man in Search of a Soul*, 122, 183, 123, 220.

28.   Swimme and Berry, *The Universe Story* (1992), 228–230.

29.   Bruchac, *Survival This Way* (1987), 333.

30.   Swimme and Berry, *The Universe Story*, 23.

31.   McKim, "The Cosmos According to Carl Sagan: Review and Critique" (1993), 24.

32.   Johnson, "'In The Beginning . . .' I Think There Was a Big Bang" (1994), quoting Sagan's *Cosmos*, 281; 48.

33.   McDaniel, *Earth, Sky, Gods & Mortals* (1990), 6.

34.   Stanton and Guernsey, "Christians' Ecological Responsibility: A Theological Introduction and Challenge" (1993), 2–3.

35.   Ibid., 3–5, 7. Because few individuals are cautious in the distinctions between ecology and environment or ecologist and environmentalist, it is not entirely clear that these authors *meant* ecologist—the scientist. Still, there is reason to believe each of us is an ecologist in the sense of being capable of appreciating ecosystems on some personally profound level.

36.   Polkinghorne, "Cross-Traffic Between Science and Theology" (1991), 145. For an expansion beyond Christianity and thus relevant to culture at large, see Matthew Fox, *The Coming of the Cosmic Christ*, 1988, New York: HarperCollins.

37. McDaniel, *Earth, Sky, Gods & Mortals*, 127–128. "Panentheism," coined some two hundred years ago, is distinguished from pantheism and means "everything in God," reflecting creation and its processes in (or by means of) God even though He is transcendent. Ibid., 137.

38. Ibid., 138.

39. Walters, "Democracy of Caring" (1996), 75.

40. Roszak, *The Voice of the Earth*, 259.

41. Orr, *Ecological Literacy* (1992), 75, 79–80.

42. Rieser, "Ecological Preservation as a Public Property Right: An Emerging Doctrine in Search of a Theory" (1991), 395–396.

43. Tuan, *Passing Strange and Wonderful* (1993), 200.

44. Rothenberg, "Individual or Community?" in Chapple (1994), 83.

18

Toward a National Environmental Ethic

The foregoing chapters have been an extended argument for a more effective environmental ethic to motivate this nation's policy and law. The case has been made for engaging the individual through instilling or evoking his or her own personal environmental ethic, informed by cultural elements ranging from the sense of place to surrogates in the archetypes found in popular culture. But for any societal turnaround to occur, the institutions of policy and law themselves must be receptive to a seemingly innovative approach. In this concluding chapter those institutions are placed in the contexts already explored throughout the preceding chapters, in search of justification for receiving the so-called new stories underlying the calls for a renewed connection with the natural world.

Those new stories are being told in science, in a proffered worldview, in a restored relationship between the material and the spiritual, and in the interconnectedness among all beings — human, biological, geological. The remaining question is whether any new story is to be found in public or private institutions of policy formulation — or whether they are prepared to advance the narratives found elsewhere.

EFFECTIVE INDIVIDUAL ARTICULATION
OF A PERSONAL AESTHETIC

It stands to reason that decisions affecting natural resources must be examined in the context of contemporaneous values. Indeed, the premise here is that those contemporaneous values are now emergent among individuals and will serve in the articulation of a new national environmental ethic. Because the decades of the environment were inspired by events at the turn of this century and are occurring in its second half, 1900 is something of a benchmark.[1]

At that time, with per capita income a tenth of what it is now, most of the population was in no financial position to be demanding aesthetic values. Leisure activities did not include the backpacking, canoeing, fishing, or hunting so popular now. Instead, those activities served as transportation and food production. And that untamed wilderness to which we are turning for the respite of peace and quiet was deemed a dangerous nuisance.[2]

Today we find it necessary to relearn how to love both nature (or wilderness) and ourselves and to resolve the array of conflicts in the name of fundamental harmony. What we should be seeking is to protect both ourselves and the natural world from our own misbehavior and to develop (reestablish) creative, harmonious relationships.[3] There is the inevitable caveat here: if our efforts are myopic and unsound, if the approach is a cynical manipulation of the concepts and values expressed here and of the emergent worldview professed by visionaries and pragmatists alike, we will fail. Our guide to addressing environmental problems will be neither realistic nor adequate.[4]

We could do worse than to turn to childhood for inspiration. Children are endowed with a sense of wonder to which we should be paying attention, not merely lip service. Children have an open ability to learn and know and an "innate need and positive desire . . . to create" to those ends. Those capacities themselves suggest a "biological basis for the attitude of wonder as we have experienced it personally, in 'the first poetic spirit of our life.'"[5]

Another source explored here is the noble savage, an archetypal internality not so far removed from the child in us. Hoxie Neale Fairchild, the chronicler of the archetypal noble savage is worthy of another look. Fairchild speaks of the profound appeal of "unintellectual activities," the delight of "the free play of instincts and their sublimations in sentiment and emotion." He appreciates literature and art that represent the addition of "the creative intelligence of a rational" person "to the eye and heart of a child." He is not indifferent to "wild spontaneity" of the "least classical genius." His emotions are normally active whether playful or solemn.[6]

And yet Fairchild speaks of the constant overlap of thinking and feeling in accepting a "real difference" in reason's willingness to think and feeling's willingness to believe, which "may instinctively be impelled to form the same hypothesis." With regard to civilization, he notes the amassing of "useless objects, formulas and conventions" bearing no relationship to our "real needs." "When these parasitic growths threaten the life of the tree, they must be torn away." Further, it is good for our "pride to know that the scenes which [we have] influenced least are often the most beautiful, that riches are not the proper goal of human effort, and that an intelligent peasant may be wiser than a stupid scholar."[7]

And thus, Fairchild observes that throughout time, the noble savage, like archetypes, comes whenever he is needed. His virtues, however portrayed, help us "to see the superficiality of those ideals of polish, worldly power, luxury, common sense and elegant learning that they had been holding before them."[8] What his chronicle illustrates is that the noble savage also takes the form best suited to the civilization giving rise to him. The same fortuity is attributed to the Green Man. And, notably, the treasure of the noble savage's close kin, the Wild Man, is not always monetary.

From our childhood and the imagination in which the images of the collective unconscious reside, we turn to education. Formal education in this nation has been the foundation serving a public prepared for an active role in government. Indeed, a precept of our Constitution is that the people are sovereign and the government is there to serve the sovereign. A stupid or ignorant sovereign is a hazard to the future — of the nation, at the very least. And so, educational institutions willing to tell the new stories are essential. Systematic surveys may well indicate that such institutions are growing in number. Two programs will represent them here, reported in 1994 in *Science* and *The Wall Street Journal*, respectively.

The first is University of California at Berkeley's College of Natural Resources. In what is described as a major renovation, the "faculty tore down old disciplinary walls and regrouped into new, broader departments. The express goal was to "bring ecological perspectives to bear on [societal] agricultural and resource issues." Accompanying this administrative alteration is a shifting of research from traditionally narrow topics toward broader questions of resources as holistic entities. Faculty and students are actually working between disciplines. Comparable movements at Midwest institutions are described in the feature despite expected repercussions in granting and faculty positions and the inevitable problems of internal politics. The fear is that the Berkeley program may not survive the twentieth century.[9]

A new synthesis, if not a new story as such, is to be found in the collaboration of New York University and the New York Botanical Garden. While the mission is the pragmatic one of discovering and exploiting plants of economic and medicinal value, the means go beyond botanical sciences. An innovative doctoral program provides training in anthropology and international law as well.[10] One can only hope that the integration proceeds to a cooperative effort serving all interests through this array of disciplines, rather than advancing exploitation in the derogatory sense that word too often conjures.

That there is hope to be found in institutions of postgraduate education and their students is suggested in the 1993 unpublished internship journal of a candidate for Vermont Law School's Master of Studies in [Environmental] Law.[11] Working with a First Nation the intern reported what subsistence meant to the elders recalling their childhoods: working only a few days a week and passing time in festivities and spiritual aspects of their lives. The intern came to the realization that this people lived well and comfortably without being wasteful. An irony was reported as well: upon decimating a fish population, Europeans declined to seek from this aboriginal people any advice on how to restore the population and instead imposed arbitrary regulations on the very ones who were once the population's caretakers. After only a summer's brief contacts with this people, the intern was calling for the consideration of the spiritual in addressing environmental issues. He knew that the stories of resources are not to be relegated to some quaint custom or some hidebound, empty religion as too often conceived by our culture. This intern discovered for himself that religion for this people is a vital context for life. It could not have been a surprise that the people have no words for either boss and management or religion and environment. The relationships are innately a part of their psyches.

EXPRESSION AT THE NATIONAL LEVEL

Legal Foundations

In the recent decades of the environment commencing in the 1960s, through passage of the Wilderness Act in 1964 and legislation into the mid-1970s Congress has created a national system protecting wildlife, free-flowing rivers, and wilderness trails. With the addition of 1969's National Environmental Policy Act, the federal government has reflected an increasing will to forgo at least "some marginal economic enterprise for the sake of preserving a sample of our heritage." Unfortunately, such a turnaround here and elsewhere seems always to be "a function of the increasing scarcity, and thus increasing value, of natural environments." But at the same time it is "a tribute to the countless individuals who have worked in the sometimes thankless task of representing noneconomic values to the nation's decision makers."[12] Thus, there is the prospect of even more profound changes of heart in Congress.

Wilderness and wild species are an important aspect of this nation's environmental protection policy, but the policy is broader than that, if not necessarily as appealing to the imagination. Pollution control and environmental protection are the counterparts. Some of this body of law finds a portion of its own foundation in the public trust doctrine.

Basically, this ancient doctrine with its roots deep in English history holds that the sovereign holds certain rights in the ground underlying navigable waters in trust for the common use of the people. Any grant of any right purporting to divest the people of those common rights is void. This form of public trust has passed to the several states without the intervention of the Constitution. (The public trust in federal lands is statutory, not a descendant of the ancient form.) Until recent times our version of the public trust has been continuously limited to lands associated with navigable waters. But there are those who would extend the doctrine on behalf of other natural features.[13]

But even if public trust was to be extended, it can hardly preclude the difficult task of setting priorities among public trust uses. Aesthetics are only one of any number of issues in this regard, including just who the public is with regard to any given natural resource, as those involved in Western water allocation controversies can attest. At least some public trust uses of such resources can be mutually exclusive.[14]

Public trust scholar Joseph L. Sax traces the doctrine to the Roman concept of common properties, *res communis*. Not only is alienation of such property or rights in it at least constrained, there is also a fiduciary's affirmative duty of protection. *Res communis belonged* to no one. It was neither amenable to possession nor subject to purchase, sale, or exclusion. Sax identifies as the central idea of public trust the prevention of "the destabilizing disappointment of expectations held in common but without formal recognition such as title," including protection against destabilizing changes. This certainly sounds like the sort of protection to be afforded the natural world, especially those portions of it not adequately protected by existing environmental policy and law. Moreover, it could be applied to anything from ownership of private property to an ecosystem in need of protection. An intriguing element is that the fact of the expectation has greater credence than the imposition of any formality.[15]

There are, then, indications of some grounding for a new environmental ethic. But the law seeks models — precedent — for political justification of innovation.

Political Justification for an Expanded Vision for Environmental Policy

We have a tendency to take ideas of justice for granted through unexamined, even unconscious assumptions. For a new sense of justice serving the natural world to take form, it must be raised to a conscious level. This book and all those other works cited and beyond citation here are certainly bringing the natural world and our interrelationship to our attention. Now, presumably, we are "free to modify, accept, or reject" the related ideas of justice. In doing so, it is wise to articulate theories to organize our principles and set priorities in an effort to avoid confusion.[16]

One view is of a utopian environmental policy, one which extends beyond our present body of law to include recognition of our obligations to future generations to provide a world still marked by "special places and natural resources" affording a quality of life at least equal to our own. While a utopian view may be unattainable, it can inform contemporary policy and aid in the move away from the impedimenta of command-and-control correction of the marketplace.[17] A hopeful omen here is the implicit change in the thrust of the phrase "quality of life."

Thomas Berry suggests a curriculum that might serve such a utopian prospect. He suggests teaching through the phases in cultural development of our own kind, from the paleolithic to the neolithic village to the time of great religious cultures to the phase of science and technology to the emergent ecological phase. Such a curriculum should assist all the participants in "envisioning the historical mission of the times" and in providing "meaning to life that might not otherwise be available." According to Berry, in this new ecological context, law "would function with a greater sense of the inherent rights of natural realities."[18]

Others suggest a similar return to abounding multi-cultural creation stories. All of them open understanding to diverse worldviews and to traditions and their arts. They also teach what we aspire to as people, our intended legacy to our children.[19] Still others, represented this time by a lawyer — in a scholarly article suggesting both the isolation of law from the larger world and the ubiquitous nature of the urge for some *more*, urge our own mythologizing of the earth. The premise is that the vacuum in environmental law is in need of an ecophilosophy to be filled with a myth or story. Gaia is a theme that fuses science with narrative, helping even lawyers perceive the human place in the biosphere and reminding this observer and us that neither science nor story alone moves us; the fusion is essential.[20]

Yet another lawyer reminds us that it is only "moments of historical crisis" or with the entry of theories from beyond law that the law will question "prevailing conventions." It is a role of legal scholarship to do the questioning. Remarking on the "lack of institutional inspiration" in the drudgery manifested through the bureaucracy of contemporary environmental policy and law, Richard O. Brooks calls for a move from concentration on symptoms to a focus on fundamental causes. What Brooks proffers is "articulation of a naturally-based philosophy of law." The attendant ideals are holding the environment in trust for future generations, respect for non-human nature, protection of vulnerable persons, and protection of natural beauty.[21]

Significantly, while such ideals may be discussed in the literature of scholars and professionals, "this ethical analysis has often been hermetically separated from a historical and legal study of the operation of these ideals and the ways in which these ideals are institutionalized in everyday practice." Brooks infers hope for a change in the increasing involvement of citizens through citizen suits, land trusts, local community activities, and even the recently popular privatization of government services, some in the name of environmental protection. One requisite to be met is to allow students to grasp ecological ideals along with the nitty-gritty details[22] — or to allow them to articulate, share, and challenge their own!

Along the way it would behoove us to look beyond our own culture and its gargantuan politico-legal system, which too often moves with the finesse and speed of a slug. While we may look on single-issue political action with some suspicion, there are the green political parties of Germany and Australia. The former is a coalition moving beyond the ordinary political party to combine individual effort with political activism, with ecology (environmentalism?) first on the platform. The latter fosters biocentrism through a mixture of ecological consciousness and attention to public policy issues.[23]

The constitutions of a number of the states are certainly closer to home and provide a variety of models for placing environmental matters high in the hierarchy of policy and law. Just as close to home, at least in some senses, are the models to be found among our own First Nations. We have appropriated much from them. Is it not possible to learn from the wealth of their experience without *appropriating* still more? Inspiration may serve the source as well as the recipient.

Just how much our unique constitutional democratic republic actually borrowed from First Nations is controversial, but the suggestive impression remains in the very fact that the Constitution itself represents a unique system of sovereignty, government, and freedom. Our political roots may be English but to what extent have they mingled with those of the First Nations the founders encountered? The question is not academic because incorporation of novel concepts into our law and policy is far more likely if precedent can be found for doing so. That First Nation ideals, if not their philosophies, found their way into the Constitution provides the precedent for formally embracing more of those ideals today.

To begin with the concept of freedom, western literature suggests it is in reference to a political entity and from domination by another entity. For us it is found in personal liberty as well as national sovereignty. The individual has passed from slavery to freedom to become a person rather than property. And in fact, our ancestors on this continent were amazed by the form of personal liberty they found here in the individual Amerindian's freedom from both rulers and social classes based on ownership. Credence is thus lent the argument that our contemporary version of democracy and liberty holds less of classical Greece and Rome or of England or France and more of "Indian notions translated into European language and culture."[24] According to Henry Steele Commager "Europe only imagined the Enlightenment, but America enacted it." A legitimate question is why.

One equally legitimate answer is suggested in the form of "a unique blend of European and Indian political ideas and institutions along the Atlantic coast between 1607 and 1776." According to Weatherford, that Spain, France, Holland, and Britain have not devised a real counterpart is a further indication of unique influence in the establishment of this nation. There is certainly ample circumstantial evidence in support of at least a subliminal contribution in the minds of the founders as they went about the lengthy process of devising what became our founding document.[25]

Benjamin Franklin[26] is a case in point, especially when identified as "the savage as philosopher." Here was a profoundly adaptable pioneer of science, politics, and morals. His natural sagacity anticipates "the traits later decades would associat with the frontier hero" — not to mention Burroughs's Tarzan of the Apes! But, Tarzan (at least of the Weissmuller ilk), certain activists in the environmental movement, and contemporary loner-heroes, like "the rugged, uncouth backwoodsman," may be too extreme to "identify with Franklin because the latter's naturalness is always domesticated with a morality as seemingly simple as it is durable, while being attuned to a milieu marked by substantial refinement."[27] Interestingly enough, this is also a description of our surrogate Tarzan as Burroughs originally developed the character.

Franklin, then, would seem to be the perfect nonfictional model for ourselves and for our environmental policy as well as for the nation he helped found. As a youth he was unsophisticated, hardworking, and rustic, yet commercially aware and quite cognizant of the means to personal success. His endeavors and virtues elevate his community along with self. With maturity he becomes aware and worldly wise. He is subtle but independent through his own industry, frugality, and diligence. Through his independence he is freed to contribute to the larger society.

Franklin is a kind of noble savage *cum* Wild Man, not unlike Tarzan and his remote descendant Mr. Spock. Franklin and these archetypal figures manage "to mediate between sophistication and simplicity, to balance and harmonize self and others by satisfying the demands of both individual and the group." In surprisingly comparable senses, they all reconcile stability with progressive change.[28]

And "Franklin stands out as a factual yet ideal hero of an urban middle landscape. His world is a careful blend of myth and reality." Moreover, once repudiating the city of Boston, he reconciles with it and "becomes intellectually indebted to the city of his birth."[29] Significantly, Benjamin Franklin was also a student of Indian assemblies and served as Pennsylvania's Indian commissioner. He became a champion of Indian political structure and advocated emulation of the Iroquois League.[30]

It would appear that the Iroquois Confederacy is the precedent we are seeking. From a contemporary perspective the Confederacy is credited with a system of checks and balances, with popular participation in decisions and direct representation along with the equivalents of states' rights and even bicameral legislatures.[31]

There is a contemporaneous account against which to compare our modern perspective, and it is available in facsimile. Lewis Henry Morgan, distressed by the displacement of First Nations and seeking to provide a "truer knowledge of [Iroquois] civil and domestic institutions," strove to describe this people in their own terms.[32] Morgan believes the Confederacy (or League) was formed early in the sixteenth century, originally to resist contiguous peoples. What he found in the League was a federal system, wisdom of civil institutions, and sagacity in administration. To his knowledge no other First Nation east of the Mississippi "had reached such a position of authority and influence, or were bound together by such enduring institutions." He speaks of the League's "friendly relations" with first the Dutch and then the English in a "covenant chain" that the Iroquois honored "with singular fidelity" until the independence of the United States "terminated the jurisdiction of the English." In the 1783 treaty between the United States and Great Britain there was no provision for the Iroquois, abandoned by Great Britain.[33]

Significantly, the League was impeded by its own hunter mentality but strove to counter its influence through "adjusting the confederacy . . . as a political fabric composed of independent parts . . . adapted to the hunter state" and yet containing "the elements of an energetic government." Morgan reports that the League nations fell suddenly and simultaneously from some internal or external perturbation and might have vanished without a trace, because "[a] verbal language, a people without a city, a government without a record, are as fleeting as the deer and the wild fowl."[34]

The legacy they gave or left us includes institutions based on elaborating family relationships moving from the individual to the tribe to the nation to the League arising from "one protracted effort of legislation" and remaining relatively stable over generations. The phenomenal detail Morgan provides is not immediately relevant here, but highlights are well worth noting: executive, legislative, judicial authority; annual legislative sessions "for the common welfare"; extraordinary sessions; international relations; distribution of duties among the sachems; the influence of all individuals, men and women, taken into account; election of chiefs as a reward for merit; military power in war-chiefs, but not encouraged by the sachems; each nation having its own keepers of the faith; one people, one government, one executive, but not so centralized that "national independencies disappeared." The great mission of this League? Peace; "to break up the spirit of perpetual warfare," described by Morgan as "the highest and the noblest aspect in which human institutions can be viewed."[35]

Morgan has no reason to report on the relationship of the members of the Iroquois League to the natural world, but almost certainly, we could learn from that aspect of their lives as well.

And so the models are there for us as individuals and as a culture and nation. What, then, are the impediments to turning to them and to embracing them and our own emergent individual environmental ethics in promoting an effective new national environmental ethic for policy and law? Only the most obvious will be noted here. We can all make our own detailed lists.

Impediments

Turning again to Thomas Berry, we find a statement of a major impediment in the very formal education upon which our system depends. Berry urges that education should be a pervasive experience and should be transformed into an integrating factor, because we need the historical continuity it can provide. College represents an opportunity for reflection on meaning and values and the creation of visions. Students should be immersed in significant historical and personal processes. In the current vernacular — "*not!*"[36] In older slang — "not bloody likely!"

Turning to environmental law and policy, we meet not integration, but one or another struggle akin to "shotgun weddings" between law and other professionals. We are in fact embroiled in a transitional interval with continuing "significant impact on the business, legal, and economic communities. Natural resources that once seemed inexhaustible now are clearly finite and in some cases scarce. These resources that once were only valued when reduced to possession now have an existence value in the wild." All three branches of government have now demonstrated "their commitment to this change. The business community must recognize the change and live in harmony with a new set of priorities that society has set."[37]

Economic value remains a guiding principle, but it could be harnessed for a more positive role in protecting environmental amenities. Certainly, massive subsidies in any one usage of a resource suggest that resource itself to be inexpensive — to lack value.[38] An associated caveat is in the twin admonitions that priceless does not mean worthless and our inability to set a price on aesthetic or spiritual value does not place those considerations in a portion of the natural world below all others.

Our wise Amerindian colleagues suggest we may be back at our earliest beginnings "where we should be. Before we make any move we should look back at where we come from so we don't step in the cow pad."[39]

And then there is the grim matter of relations between the United States and the First Nations. It seems we are neither wiser nor more equitable today than we have been since the first Amerindian was forced away from the land of ancient wanderings. How can we learn from a people for whom we show so little respect?

Three cases are entirely too representative. *Shoshone Indians v. U.S.,* 324 U.S. 335, 65 S.Ct. 690, 89 L.Ed. 985 (1945) addressing the foreign concept of title in land: title was held to have been transferred from the tribe, despite, as a concurring justice put it: "People do not have words to fit ideas that have never occurred to them. Ownership meant no more to them than to roam the land as a great common, and to possess and enjoy it in the same way that they possessed and enjoyed sunlight and the west wind and the feel of spring in the air. Acquisitiveness, which develops a law of real property, is an accomplishment only of the 'civilized'" (324 U.S. 357). *Badoni v. Higginson,* 455 F. Supp. 641 (1977): a claim for correction of destruction of holy places and to preclude further destruction and desecration failed *for lack of a property interest in those places of the gods* because it was never a part of the reservation and was owned by the federal government as a portion of the National Park System. And that was just the beginning of the insensitivity expressed by the court. *Sequoyah v Tennessee Valley Authority,* 620 F2d 1159 (1980): injunction against the completion of Tellico Dam denied and action dismissed because the claims raised, according in part to anthropologists, did not reflect that the places at risk represented the cornerstone of religious observation.

And yet there may be hope here too. The U.S. Environmental Protection Agency is demonstrating a level of respect not apparent in the three levels of the federal judicial system. The agency reportedly approved water quality standards established by a New Mexican pueblo. The standards were more stringent than either their federal or their state counterparts. In the process the agency demonstrated respect for the Amerindian value judgment and corresponding protections. Indeed, the larger issue here is that "[f]or the Native American reverence for the life and spiritual nature of all of creation to be truly meaningful, it must receive legal recognition within tribal court jurisprudence. If it does not, the observation embodies nothing more than a romantic cliche."[40] Additional good news here is that the district court upheld the Environmental Protection Agency's approval.

Important in this and the larger context is the role of tribal narrative, the "corpus of cultural stores which incorporate important traditional values and elements which bear on larger matters of tribal history, identity, and destiny." The challenge Frank Pommersheim poses is finding the best way "to move away from an emphasis on polarization, division, stratification and towards conciliation, unity, and respect." It is best done in small steps.[41] Certainly, there are circles within circles and intermeshing spirals — if only we can discern and act upon them.

Pommersheim himself notes that all the legal struggles in Indian country, as it is often called these days, "over land, water, natural resources, and jurisdiction [are] a fervent commitment to place." It is Amerindian tribal courts who hold the potential "to blend the best of both worlds. Such a synthesis would recognize the importance of language, of narrative, and of story, as well as the meaning of justice from the indigenous point of view."[42] Without any flippancy at all, we should live so long!

The final impediment to be identified here is our relationship to the legal system itself. To put it in the briefest of terms: "Because voters are rationally ignorant, because benefits can be concentrated and costs diffused, and because individual voters seldom (and probably never) influence the outcome of elections, there is little reason to expect that elections [for one] will link political decisions to efficiency,"[43] whatever the context and however defined.

In this context it is truly daunting to discover that in the experience of the Hetch Hetchy Valley damming, the outcry for environmental amenities was "proportional to the distance from it." A comparable situation developed in the 1970s when Alaskan land was at risk and hearings were held at varying distances from the wilderness in question. As hearings progressed with increasing distance from Native villages to settled towns to urban Alaska to the American West to the American East, favorable attitudes toward the wilderness progressed in parallel from zero percent in stages to 100 percent.[44] It is indeed time for a turnaround!

THE CALL FOR PERSONAL COMMITMENT TO AN ENLIGHTENED ENVIRONMENTAL ETHIC, STEWARDSHIP OF THE EARTH

No one can or should dismiss freedom, "liberation from oppression and achievement of equal opportunity"; it "is a prerequisite for full human self-realization, but freedom by itself does not assure well-being and fulfillment." Instead our freedom confronts us "with a supreme spiritual challenge." In a free world we are called upon as individuals to perfect our "freedom though faith and responsible action. Freedom without guidance by wisdom is dangerous. It easily becomes destructive to the individual and to the community, or communities, upon which every individual is dependent."[45]

With regard to wilderness, standing representative of environments and the natural world, its ultimate future "will depend less on the partisan control of the Congress or presidency, and more on the tides of public sentiment to which both parties inevitably, if sluggishly, react."

Americans seem torn between two competing concepts of our place on the planet. We are conditioned by our past successes to believe in the technological fix: This school of thought presumes that as every resource is exhausted, it will be replaced by some new wonder of technology. This view supports the argument that oil conservation is unnecessary, because we have plenty to last us until the advent of breeder reactors and controlled nuclear fusion. In this sort of thinking, wilderness appreciation can be replaced by movies and television and ultimately by pleasure electrodes in our heads, and we will all be the happier for it.[46]

In a bit of contemporary history, one can only hope to be prophetic, President Clinton has admitted to First Nation leaders that "[s]o much of who we are today comes from who you have been for a long time. . . . Because of your ancestors, democracy existed here long before the Constitution was drafted and ratified. . . . I believe in your rich heritage and in our common heritage." With regard to the Great Law of the Iroquois Confederacy, the President cited certain advice: "In our every deliberation, we must consider the impact of our decision on the next seven generations" and went on to assert "We are stewards, we are care-takers. That standard will keep us great if we have the vision of your fore-fathers."[47]

We are in process of "expanding our moral sensitivity beyond suicide, homicide, and genocide to include biocide and geocide, evils that were not recognized in our civilizational traditions until recently." "Until we begin to think about our human story as integral with the larger life story and the larger Earth story we will not be fully into the Ecozoic period. We will not have an Ecozoic governance."[48]

We can only hope that the time has truly come when "[t]he global discussion of environmental issues has revealed that there are many diverse cultural pathways that lead to a faith in an ethics consistent with ecology and social liberation. . . . A global environmental philosophy is now emerging."[49] Perhaps it will lead to the center of government in ecosystem instead of nation-state,[50] an intriguing prospect on any number of levels.

NOTES

1. A chronology of corresponding national policy and law in societal context can be compiled from White, *"It's Your Misfortune and None of My Own"* (1991); Doron, *Legislating for the Wilderness* (1986); Allin, *The Politics of Wilderness Preservation* (1982); and Anderson and Leal, *Free Market Environmentalism* (1991).

2. Anderson and Leal, *Free Market Environmentalism*, 45.

3. Tucker, "An Ecological Cosmology" in Chapple (1982), 150–152.

4. Kealey, *Revisioning Environmental Ethics* (1990), xii.

5. Cobb, *The Ecology of Imagination in Childhood* (1977), 24.

6.   Fairchild, *The Noble Savage* (1928), 500.

7.   Ibid., 501, 508.

8.   Ibid., 510.

9.   Barinaga, "A Bold New Program at Berkeley Runs into Trouble" (1994), 1367–1368. However, the environmental programs of Antioch New England Graduate School are evidence of promising future programs.

10.   Anonymous, *The Wall Street Journal*, October 20, 1994, A1.

11.   Presented to the author in her capacity as the Environmental Law Center's Director of Internships.

12.   Allin, *Politics of Wilderness Preservation*, 178.

13.   Stevens, "The Public Trust: A Sovereign's Ancient Prerogative Becomes the People's Environmental Right" (1980), 195–196, 199. The doctrine was either conveyed or articulated in the Magna Carta, ibid, 200 and N. 28, 200.

14.   Ibid., 223, 224.

15.   Sax, "Liberating the Public Trust Doctrine from Its Historical Shackles" (1980), 186, 188, 192–193.

16.   Wenz, *Environmental Justice* (1988), 2, 43.

17.   Deutsch, "Setting Priorities: Principles to Improve Environmental Policy," 68, 43–45.

18.   Berry, *The Dream of the Earth* (1988), 101, 104.

19.   Hill and Hill, eds., "Creation's Journey" in Hill and Hill (1994) 30, excerpts from inaugural exhibit and book commemorating the opening of the National Museum of the American Indian in New York City.

20.   Kwisnek, "Comment: Earth or Consequences? Mythologizing the Earth Entity as a Way to Environmental Awareness" (1991), 737, 739, 741, 742.

21.   "A New Agenda for Modern Environmental Law" (1991), 1, N. 1 on 1, 2, 11, 13, 14.

22.   Ibid., 15, 20–21, 23.

23.   Devall and Sessions, *Deep Ecology* (1987), 9.

24.   Weatherford, *Indian Givers* (1988), 128.

25.   Ibid., 134–136.

26.   Other founders influenced by First Nations include James Madison and Thomas Jefferson, who also were influenced by the Iroquois, and John Adams, who with his family socialized with Cayuga chiefs. Mander, *In the Absence of the Sacred* (1991), 230, 233.

27.   Machor, *Urban Ideals and the Symbolic Landscape of America* (1987), 116.

28.   Ibid., 118.

29.   Ibid., 118–119.

30.   Weatherford, *Indian Givers*, 136.

31.   Mander, *In the Absence of the Sacred*, 200.

32.   Morgan, *League of the Iroquois* (1993), ix, xviii.

33.   Ibid., 4, 8, 7, 10, 29.

34.   Ibid., 17, 57–59.

35. Ibid., 63, 66–67, 71, 72, 75, 77, 92.

36. Berry, *The Dream of the Earth*, 97.

37. Campbell, "Economic Valuation of Injury to Natural Resources" (1992), 29, 57.

38. Anderson and Leal, *Free Market Envrionmentalism*, 101.

39. Bruchac, *Survival This Way* (1987), 29.

40. Duthu, "Holding a Great Vision: Engaging the Jurisprudential Voice of Tribal Courts," review of *Braid of Feathers: American Indian Law and Contemporary Tribal Life* by Frank Pommersheim, 1136, 1137, quoting from Pommersheim.

41. Ibid., 1146, 1150.

42. Pommersheim, front and back flaps.

43. Anderson and Leal, *Free Market Environmentalism*, 16.

44. Allin, *Politics of Wilderness Preservation*, 222–223.

45. Rockefeller and Elder, *Spirit and Nature* (1992), 2.

46. Allin, *Politics of Wilderness Preservation*, 214, 275.

47. Guest essay, *Native Peoples* 7 (Summer 1994), 6.

48. Swimme and Berry, *The Universe Story* (1992), 257, 260.

49. Rockefeller and Elder, *Spirit and Nature*, 10.

50. Nollman, *Spiritual Ecology* (1990), 103.

Bibliography

Adams, Val. "Tarzan Coming to TV without Jane." Reprinted in *The Gridley Wave* 20 (May 1966).

Aeppel, Timothy. "For This Rabbi, Saving the Earth Just Comes with The Territory." *The Wall Street Journal*, June 1, 1994, B1.

____. "Ecotherapists Explore the Green Side of Feeling Blue." *The Wall Street Journal*, August 14, 1995, B1.

Allen, Paula Gunn. "Recuerdo." In *Survival This Way: Interviews with American Indian Poets*, edited by Joseph Bruchac. 1987. Tucson: Sun Tracks and the University of Arizona.

Allin, Craig W. *The Politics of Wilderness Preservation*. 1982. Westport, Conn.: Greenwood Press.

Alpers, Antony. *Legends of the South Seas: The World of the Polynesians Seen Through Their Myths and Legends, Poetry and Art*. 1970. New York: Thomas Y. Crowell Company.

Amiotte, Arthur. "The Road to the Center." In *I Become a Part of It: Sacred Dimensions in Native American Life*, edited by D. M. Dooling and Paul Jordan-Smith. 1989. New York: Parabola Books.

Anderson, Terry L. and Donald R. Leal. *Free Market Environmentalism*. 1991. Boulder, Colo.: Westview Press.

Anderson, William. *Green Man: The Archetype of Our Oneness with the Earth*. 1990. San Francisco: HarperCollins.

Anonymous. *The Enchanted World — Fabled Lands*. 1986. Alexandria, Va.: Time-Life.

Armstrong, Karen. *A History of God: The 4,000-Year Quest of Judaism, Christianity and Islam.* 1993. New York: Alfred A. Knopf.

Aulie, Richard P. "Al-Ghazili Against Aristotle: An Unforeseen Overture to Science in Eleventh Century Baghdad." *Perspectives on Science & Christian Faith* 45 (March 1994), 26.

Bailey, J. O. *Pilgrims Through Space and Time: Trends and Patterns in Scientific and Utopian Fiction.* 1947. New York: Argus Books.

Barinaga, Marcia. "A Bold New Program of Berkeley Runs into Trouble." *Science* 263 (March 11, 1964), 1367–1368.

Berle, Peter A. A. "Toward a Grander Vision." *Audubon* (July–August 1995) 6.

Berry, Thomas. *The Dream of the Earth.* 1988. San Francisco: Sierra Club Books.

Birrell, Francis. "The Glories of Excess." *The New Statesman and Nation*, May 21, 1932, 661.

Blier, Suzanne Preston. "The Place Where Votun Was Born." *Natural History* (1995), 40, 42, 44, 47.

Blum, Ann Shelby. *Picturing Nature: American Nineteenth-Century Zoological Illustration.* 1993. Princeton, N.J.: Princeton University Press.

Boorstin, Daniel J. *The Creators: A History of Heroes of the Imagination.* 1992. New York: Random House.

Botkin, Daniel B. "Ecological Theory and Natural Resource Management: Scientific Principles or Cultural Heritage?" In *Ecological Prospects: Scientific, Religious, and Aesthetic Perspectives*, edited by Christopher Key Chapple. 1994. Albany: State University of New York Press.

Boyum, Joy Gould. "New Films: Stevie and John, Arthur and Tarzan." *The Wall Street Journal*, 1981.

Brackett, Leigh. *The Secret of Sinharat.* 1964. New York: Ace Double Edition.

____. *People of the Talisman.* 1964. New York: Ace Double Edition.

Brandenburg, Jim. *Brother Wolf: A Forgotten Promise.* 1993. Minocqua, Wis.: NorthWorld Press, Inc.

Broadhurst, Paul. "The Subtle Power of Myth." *Orion* (Summer 1995), 12.

Brooks, Richard O. "A New Agenda for Modern Environmental Law." *Journal of Environmental Law & Litigation* 6 (1991), 1.

Brooks, Tim and Earl Marsh. *The Complete Directory to Prime Time Network TV Shows 1946–Present.* 1979. New York: Ballantine Books.

Brown, Joseph Epes. "The Bison and the Moth." In *I Become a Part of It: Sacred Dimensions in Native American Life*, edited by D. M. Dooling and Paul Jordan-Smith. 1989. New York: Parabola Books.

Bruchac, Joseph. *Survival This Way: Interviews with American Indian Poets.* 1987. Tucson: Sun Tracks and The University of Arizona.

Brunner, David L., Will Miller, and Nan Stockholm, eds. *Corporations and the Environment: How Should Decisions Be Made?* 1981. Stanford, Calif.: Committee on Corporate Responsibility, Graduate School of Business, Stanford University

Buel, Ronald. "The Swinger's Return: Tarzan (but not Jane) in a Jungle Comeback." *The Wall Street Journal*, 1966.

Burroughs, Edgar Rice. *Tarzan and the Castaways*. 1964. New York: Ballantine Books.

_____. *Tarzan the Magnificent*. 1964. New York: Ballantine Books.

_____. *Tarzan and the Madman*. 1964. New York: Ballantine Books.

_____. *The Moon Men*. 1960. New York: Ace.

_____. *The Eternal Savage*. 1960. New York: Ace edition of *The Eternal Lover*. 1914. New York: Frank A. Munsey Co.

_____. *Tarzan and the "Foreign Legion."* 1947. Tarzana: ERB, Inc.

_____. *Tarzan and the City of Gold*. 1938. Tarzana: ERB, Inc.

_____. *Tarzan and the Forbidden City*. 1938. Tarzana: ERB, Inc.

_____. *Tarzan's Quest*. 1936. Tarzana: ERB, Inc.

_____. *Tarzan and the Leopard Men*. 1935. Tarzana: ERB, Inc.

_____. *Tarzan and the Lion Man*. 1934. New York: Grosset & Dunlap.

_____. *Tarzan the Invincible*. 1930–1931. New York: Grosset & Dunlap.

_____. *Tarzan at the Earth's Core*. 1929–1930. New York: Grosset & Dunlap.

_____. *Tarzan and the Lost Empire*. 1928–1929. New York: Metropolitan.

_____. *Tarzan, Lord of the Jungle*. 1928. New York: Grosset & Dunlap.

_____. *Tarzan and the Ant-Men*. 1924. New York: Grosset & Dunlap.

_____. *Tarzan and the Golden Lion*. 1923. New York: Grosset & Dunlap.

_____. *Tarzan the Terrible*. 1921. New York: Grosset & Dunlap.

_____. *Tarzan the Untamed*. 1920. New York: Grosset & Dunlap.

_____. *Jungle Tales of Tarzan*. 1919, 1963. New York: Ballantine Books.

_____. *Tarzan and the Jewels of Opar*. 1918. New York: Grosset & Dunlap.

_____. *The Son of Tarzan*. 1918. New York: Burt.

_____. *The Beasts of Tarzan*. 1916. New York: Grosset & Dunlap.

_____. *The Return of Tarzan*. 1915. New York: Grosset & Dunlap.

_____. *Tarzan of the Apes*. 1914. New York: Grosset & Dunlap.

Callicott, J. Baird. "The Wilderness Idea Revisited: The Sustainable Development Alternative." In *Ecological Prospects: Scientific, Religious, and Aesthetic Perspectives*, edited by Christopher Key Chapple. 1994. Albany: State University of New York Press.

Campbell, Thomas A. "Economic Valuation of Injury to Natural Resources." *Natural Resources & Environment* 6 (Winter 1992), 29, 57.

Card, Orson Scott, *Hatrack River* [an anthology of *Seventh Son*, 1987; *Red Prophet*, 1988; *Prentice Alvin*, 1989). n.d. New York: Tom Doherty Associates.

Caulfield, Deborah. "He 'Greystoke,' They Not 'Tarzan.'" March 3, 1983.

Cavendish, Richard, ed. *Legends of the World*. 1982, 1989. New York: Barnes & Noble.

Chapple, Christopher Key, ed. *Ecological Prospects: Scientific, Religious, and Aesthetic Perspectives*. 1994. Albany: State University of New York Press.

Chester, William L. *Kioga of the Unknown Land.* 1978. New York: Daw Books.
____. *One against the Wilderness.* 1977. New York: Daw Books.
____. *Kioga of the Wilderness.* 1976. New York: Daw Books.
____. *Hawk of the Wilderness.* 1935. New York: Ace Books, Inc.
Churchill, Ward. *Fantasies of the Master Race: Literature, Cinema and the Colonization of American Indians.* M. Annette Jaimes, ed. 1992. Monroe, Me.: Common Courage Press.
Clifford, James. *The Predicament of Culture: Twentieth-Century Ethnography, Literature, and Art.* 1988. Cambridge, Mass.: Harvard University Press.
Cobb, Edith. *The Ecology of Imagination in Childhood.* 1977. New York: Columbia University Press.
Colinvaux, Paul. *Why Big Fierce Animals Are Rare: An Ecologist's Perspective.* 1978 (1979 paperback ed.). Princeton, N.J.: Princeton University Press.
Connor, George, ed. *Listening to Your Life: Daily Meditations with Frederick Buechner.* 1992. San Francisco: Harper San Francisco.
Conron, John, ed. *The American Landscape: A Critical Anthology of Prose and Poetry.* 1974. New York: Oxford University Press.
Corriel, Vern. "Tarzan the Ape Man/Me Bo — Me Show." *The Gridley Wave* 83 (Summer 1981).
Davies, Paul. *God and the New Physics.* 1983. New York: Simon & Schuster.
De Camp, L. Sprague. *Lost Continents: The Atlantis Theme in History, Science, and Literature.* 1970. New York: Dover Publications, Inc.
De Chateaubriand, Francois-René. *Atala* (1801) and *Rene* (1802). (Walter J. Cobb, trans.). 1961. New York: Signet Classic/The New American Library.
Deloria, Vine Jr. "Out of Chaos." In *I Become a Part of It: Sacred Dimensions in Native American Life,* edited by D. M. Dooling and Paul Jordan-Smith. 1989. New York: Parabola Books.
Deutsch, Stuart L. "Setting Priorities: Principles to Improve Environmental Policy." *Chicago Kent Law Review* 68, (1992): 43–45.
Devall, Tim and George Sessions. *Deep Ecology: Living as if Nature Mattered.* 1987. Layton, Ut.: Gibbs Simith Publications.
Dick, Philip K. *Do Androids Dream of Electric Sheep?* 1968. New York: Del Rey.
Dobb, Edwin. "Back to the Future." *Audubon* (September–October 1995), 80.
Dooling, D. M. and Paul Jordan-Smith, eds. *I Become Part of It: Sacred Dimensions in Native American Life.* 1989. New York: Parabola Books.
Donohue, John J. *Warrior Dreams: The Martial Arts and the American Imagination.* 1994. Westport, Conn.: Bergin & Garvey.
Doron, William D. *Legislating for the Wilderness: RARE II and the California National Forests.* 1986. New York: Associated Faculty Press, Inc.
Dorm-Adzobu, Clement, Okyeame Ampadu-Agyei, and Peter G. Veit. *Religious Beliefs and Environmental Protection: The Malshegu Sacred Grove in Northern Ghana,* edited by The Center for International Development and

Environment. 1991. Nairobi, Kenya: African Centre for Technology Studies.

Douglas, William O. *My Wilderness: East to Katahdin.* 1961. Garden City, N.Y.: Doubleday & Company, Inc.

Dudley, Edward and Maximillian E. Novak, eds. *The Wild Man Within: An Image in Western Thought from the Renaissance to Romanticism.* 1972. Pittsburgh, Pa.: University of Pittsburgh Press.

Duthu, N. Bruce. "Implicit Divestiture of Tribal Powers: Locating Legitimate Sources of Authority in Indian Country." *American Indian Law Review* 19 (1994), 353.

Eastman, Charles A. *From the Deep Woods to Civilization: Chapters in the Autobiography of an Indian.* 1916, 1944. Lincoln: University of Nebraska Press.

Elgin, Duane. *Voluntary Simplicity: Toward a Way of Life that Is Outwardly Simple, Inwardly Rich.* 1981. New York: William Morrow and Company, Inc.

Endrezze, Anita. "The Humming of Stars and Bees and Waves." In *Talking Leaves: Contemporary Native American Short Stories,* edited by Craig Leslie. 1991. New York: A Laurel Trade Paperback.

Essoe, Gabe. *Tarzan of the Movies: A Pictorial History of More than Fifty Years of Edgar Rice Burroughs' Legendary Hero.* 1968. New York: The Citadel Press.

Estés, Clarissa Pinkola. *Women Who Run with the Wolves: Myths and Stories of the Wild Woman Archetype.* 1992. New York: Ballantine Books.

Fairchild, Hoxie Neale. *The Noble Savage: A Study in Romantic Naturalism.* 1928. New York: Columbia University Press.

Fenyvesi, Charles. "The Power of Landscapes." *Washington Post,* June 21, 1995, C1.

Farney, Dennis. "Natural Questions: Chaos Theory Seeps into Ecology Debate, Stirring up a Tempest." *The Wall Street Journal,* July 11, 1994, 1.

Farmer, Philip Jose. *Tarzan Alive.* 1972. New York: Popular Library.

Feret, Bill. *Lure of the Tropix: A Pictorial History of the Tropic Temptress in Films, Serials and Comics.* 1984. New York: Proteus Books.

Ferkiss, Victor. *Nature, Technology, and Society: Cultural Roots of the Current Environmental Crisis.* 1993. New York: New York University Press.

Finch, Robert. *The Primal Place.* 1983. New York: W. W. Norton & Company.

Fontana, D. C. *The Questor Tapes.* 1974. New York: Ballantine Books.

Frome, Michael. *Promised Land: Adventures and Encounters in Wild America.* 1985. New York: William Morrow and Company, Inc.

Fury, David. *Kings of the Jungle: An Illustrated Reference to "Tarzan" on Screen and Television.* 1994. Jefferson, N.C.: McFarland & Co., Inc.

Gaddis, Vincent H. *American Indian Myths & Mysteries.* 1977. New York: Indian Head Books.

Gallagher, Winifred. *The Power of Place: How Our Surroundings Shape Our Thoughts, Emotions, and Actions.* 1993. New York: Poseidon Press.

Gardner, Gerald and Dee Caruso. *The World's Greatest Athlete.* 1973. Greenwich: Fawcett Gold Medal Books.

George, Jean Craighead. *Julie of the Wolves.* 1972. New York: Harper & Row.

Gill, Sam. "Disenchantment." In *I Become a Part of It: Sacred Dimensions in Native American Life,* edited by D. M. Dooling and Paul Jordan-Smith. 1989. New York: Parabola Books.

Glavovic, P. D. "Traditional Rights to the Land and Wilderness in South Africa." *Case Western Reserve Journal of International Law* 23 (1991), 308, 309, 318, 319–320.

Gould, Stephen Jay. *Wonderful Life: The Burgess Shale and the Nature of History.* 1989. New York: W. W. Norton & Company.

Greeley, Andrew M., *Confessions of a Parish Priest: An Autobiography.* 1986. New York: Simon and Schuster.

Grossman, Gary. *Saturday Morning TV.* 1981. New York: Delacorte Press.

Grove, Richard H. "Origins of Western Environmentalism." *Scientific American* (July 1992), 42.

Hanh, Thich Nhat. *The Miracle of Mindfulness: A Manual on Meditation,* rev. ed. Translated by Mobi Ho. 1987. Boston: Beacon Press Books.

Harjo, Joy. "Anchorage." In *Survival This Way: Interviews with American Indian Poets,* edited by Joseph Bruchac. 1987. Tucson: Sun Tracks and the University of Arizona.

Hays, Samuel P. "From Conservation to Environmentalism." In *American Environmentalism: Readings in Conservation History,* 3d ed., edited by Roderick Frazier Nash. 1990. New York: McGraw-Hill.

Heizer, Robert F. and Albert B. Elsasser. *The Natural World of the California Indians.* 1980. Berkeley: University of California Press.

Henry, Shannon. "Endangered Speech: Native Americans Attempt to Save Tribal Languages for the Next Generation." *Christian Science Monitor,* June 29, 1993, 4.

Hermundsgård, Frode. *Child of the Earth: Tarjei Vesaas and Scandinavian Primitivism.* 1989. New York: Greenwood Press.

Hill, Tom and Richard W. Hill, Sr., eds. "Creation's Journey." *Native Peoples* (Fall 1994), 30, excerpts from inaugural exhibit and book.

Hillerman, Tony. *Hillerman Country: A Journey Through the Southwest with Tony Hillerman.* 1991. New York: HarperCollins Publishers.

Hines, Stephen W., ed. *Laura Ingalls Wilder: Little House in the Ozarks.* 1991. Nashville: Thomas Nelson, Inc.

Hogan, Linda. Interview. In *Survival This Way: Interviews with American Indian Poets,* edited by Joseph Bruchac. 1987. Tucson: Sun Tracks and the University of Arizona.

Holtsmark, Erling B. *Edgar Rice Burroughs.* 1996. Boston: Twayne Publishers.

ABOUT THE AUTHOR

DOROTHY J. HOWELL, formerly an applied microbial ecologist, environmental counsel, and educator, is a candidate for the Ph.D. in Environmental Studies at Antioch New England Graduate School. She is the author of *Ecology for Environmental Professionals* (Quorum, 1994), *Scientific Literacy and Environmental Policy* (Quorum, 1992), and *Intellectual Properties and the Legal Protection of Fictional Characters* (Quorum, 1990).

ISBN 0-89789-391-3

EAN

9 780897 893916

90000>

HARDCOVER BAR CODE

NX 18Ø .S6 H72 1997
Howell, Dorothy J.
Environmental stewardship

____. *Tarzan and Tradition: Classical Myth in Popular Literature.* 1981. Westport, Conn.: Greenwood Press.

Howard, Robert E., L. Sprague de Camp, and Lin Carter. *Conan of Cimmeria.* 1969. New York: Ace Books.

Hughes, J. Donald. "Ecology and Development as Narrative Themes of World History." *Environmental History Review* 19 (1995), 1.

Itard, Jean. "The Wild Boy of Aveyron." Translated in Lucien Malson. *Wolf Children and the Problem of Human Nature.* 1972. New York: Monthly Review Press.

Jabner, Elaine. "The Spiritual Landscape." In *I Become a Part of It: Sacred Dimensions in Native American Life*, edited by D. M. Dooling and Paul Jordan-Smith. 1989. New York: Parabola Books.

Jackson, John Brinckerhoff. *A Sense of Time, a Sense of Place.* 1994. New Haven: Yale University Press.

Jamil, Tahir. *Transcendentalism in English Romantic Poetry.* 1989. New York: Vantage Press.

John Paul II. *Crossing the Threshold of Hope.* Translated by Vittorio Messori, Jenny McPhee, and Martha McPhee. 1995. New York: Alfred A. Knopf.

Johnson, Beverly Howard "'In The Beginning . . . ' I Think There Was a Big Bang." *Perspectives on Science & Christian Faith* 45 (1994), 281.

Jung, C. G. *Modern Man in Search of a Soul.* Translated by W. S. Dell and Cary F. Baynes. 1933. New York: Harcourt Brace Jovanovich.

Kaplan, Stephen. "Aesthetics, Affect, and Cognition: Environmental Preference from an Evolutionary Perspective." *Environment and Behavior* 19 (1987), 3.

Kavash, Barrie. "Native Foods of New England." In *Enduring Traditions: The Native Peoples of New England*, edited by Laurie Weinstein. 1994. Westport, Conn.: Bergin & Garvey.

Kellert, Stephen R. and Edward O. Wilson, eds. *The Biophilia Hypothesis.* 1993. Washington, D.C.: Island Press/Shearwater Books.

Kealey, Daniel A. *Revisioning Environmental Ethics.* 1990. Albany: State University of New York Press.

Kipling, Rudyard. *The Jungle Books.* 1961. New York: Signet Classics.

Kirkham, Pat and Janet Thumin, eds. *You Tarzan: Masculinity, Movies and Men.* 1993. New York: St. Martin's Press.

Kline, Otis Adelbert. *Jan of the Jungle.* 1931. New York: Ace Books, Inc.

Kwisnek, Ella A. "Comment: Earth or Consequences? Mythologizing the Earth Entity as a Way to Environmental Awareness." *Duquesne Law Review* 29 (1992), 737, 739, 741, 742.

LaChappelle, Dolores. *Sacred Land Sacred Sex — Rapture of the Deep: Concerning Deep Ecology — and Celebrating Life.* 1988. Silverton, Colo.: Finn Hill Arts.

LaFantasie, Glenn W., ed. *The Correspondence of Roger Williams, Volume I 1629–1653.* 1988. Hanover, N.H.: Brown University Press/University Press of New England.

Langdon, Philip. *A Better Place to Live: Reshaping the American Suburb.* 1994. Amherst: The University of Massachusetts Press.

Lawrence, D. H., "Escape." In *Sacred Land Sacred Sex — Rapture of the Deep: Concerning Deep Ecology — and Celebrating Life,* edited by Dolores LaChappelle. 1988. Silverton, Colo.: Finn Hill Arts.

Leopold, Aldo. *A Sand County Almanac.* 1966. New York: Ballantine Books.

Leslie, Craig, ed. *Talking Leaves: Contemporay Native American Short Stories.* 1991. New York: A Laurel Trade Paperback.

Lester, Joan. "Art for Sale: Cultural and Economic Survival." In *Enduring Traditions: The Native Peoples of New England,* edited by Laurie Weinstein. 1994. Westport, Conn.: Bergin & Garvey.

Lichtenberg, Jacqueline, Sondra Marshak, and Joan Winson. *Star Trek Lives!* 1975. New York: Bantam Books.

Lieber, Fritz. *Tarzan and the Valley of Gold.* 1966. New York: Ballantine Books, Inc.

Llywelyn, Morgan. *Finn McCool.* 1994. New York: Tom Doherty Associates, Inc.

Lopez, Barry. *Arctic Dreams: Imagination and Desire in a Northern Landscape.* 1986. New York: Charles Scribner's Sons.

Lupoff, Richard A. *Edgar Rice Burroughs: Master of Adventure.* 1965. New York: Canaveral Press.

Lyons, Oren. "Our Mother Earth." In *I Become a Part of It: Sacred Dimensions in Native American Life,* edited by D. M. Dooling and Paul Jordan-Smith. 1989. New York: Parabola Books.

Machor, James L. *Urban Ideals and the Symbolic Landscape of America.* 1987. Madison: The University of Wisconsin Press.

Maclean, Charles. *The Wolf Children.* 1977. New York: Hill and Wang.

Maloney. "Edgar of the Apes: He Had the Craziest Dream — and 'Tarzan' Was Born." *TV Guide,* October 22, 1966.

Malson, Lucien. *Wolf Children and the Problem of Human Nature.* 1972. New York: Monthly Review Press.

Mander, Jerry. *In the Absence of the Sacred: The Failure of Technology and the Survival of the Indian Nations.* 1991. San Francisco: Sierra Club Books.

Margulis, Lynn and Dorion Sagan. *Microcosmos: Four Billion Years of Evolution from our Microbial Ancestors.* 1991. New York: Touchstone Books.

Marsh, George Perkins. *Man and Nature: or, Physical Geography as Modified by Human Action,* edited by David Lowenthal. 1864, 1965. Cambridge, Mass.: The Belknap Press of Harvard University Press.

Mason, A.E.W. *The Broken Road.* 1907. New York: Charles Scribner's Sons.

Matthiessen, Peter. *In the Spirit of Crazy Horse.* 1991. New York: Viking Penguin.

Mattingly, Virginia, ed. *Songs of the Earth: A Tribute to Nature in Word and Image.* 1995. Philadelphia: Running Press.

McDaniel, Jay B. *Earth, Sky, Gods & Mortals: Developing an Ecological Spirituality.* 1990. Mystic, Conn.: Twenty-Third Publications.

____. "Emerging Options in Ecological Christianity: The New Story, The Biblical Story, and Panentheism." In *Ecological Prospects: Scientific, Religious, and Aesthetic Perspectives*, edited by Christopher Key Chapple. 1994. Albany: State University of New York Press.

McFadden, Steven. *Profiles in Wisdom: Native Elders Speak About the Earth.* 1991. Santa Fe, N.M.: Bear & Company Publishing.

McFague, Sallie. "A Square in the Quilt: One Theologian's Contribution to the Planetary Agenda." In *Spirit and Nature: Why the Environment Is a Religious Issue*, edited by Steven C. Rockefeller and John C. Elder. 1992. Boston: Beacon Press.

McGaa, Ed; Eagle Man. *Mother Earth Spirituality: Native American Paths to Healing Ourselves and Our World.* 1990. New York: HarperSanFrancisco.

McGreal, Dorothy. "The Burroughs No One Knows." *The World of Comic Art* (Fall 1965), 12, 14.

McGuane, Thomas. "The Spell of Wild Rivers." *Audubon* (November–December 1993), 62.

McKim, Mark G. "The Cosmos According to Carl Sagan: Review and Critique." *Perspectives on Science & Christian Faith* 45 (1993), 18.

McMullen, Ann. "What's Wrong with this Picture? Context, Conversion, Survival, and the Development of Regional Native Cultures and Pan-Indianism in Southeastern New England." In *Enduring Traditions: The Native Peoples of New England*, edited by Laurie Weinstein. 1994. Westport, Conn.: Bergin & Garvey.

Meeker, Joseph W. *Minding the Earth: Thinly Disguised Essays on Human Ecology.* 1988. Alameda, Calif.: The Latham Foundation.

Midgley, Mary. *Beast and Man: The Roots of Human Nature.* 1978. Ithaca, N.Y.: Cornell University Press.

Miner, Earl. "The Wild Man Through the Looking Glass." In *The Wild Man Within: An Image in Western Thought from the Renaissance to Romanticism*, edited by Dudley, Edward and Maximillian E. Novak. 1972. Pittsburgh, Pa.: University of Pittsburgh Press.

Moorcock, Michael. *The Swords Trilogy.* 1971. New York: Berkley Books.

Momaday, W. Scott. Interview. In *Survival This Way: Interviews with American Indian Poets*, edited by Joseph Bruchac. 1987. Tucson: Sun Tracks and the University of Arizona.

Morgan, Lewis Henry. *League of the Iroquois: A Classic Study of an American Indian Tribe with the Original Illustrations.* 1993 facsimile edition. New York: A Citadel Press Book.

Morgenthaler, Eric. "Design for Living: Old-style Towns Where People Walk Have Modern Backers." *The Wall Street Journal*, February 1, 1993, 1.

Mosher, Howard Frank, ed. *Songs of the North: A Sigurd Olson Reader*. 1987. New York: Penguin Books.

Nabhan, Gary Paul. "Children in Touch, Creatures in Story." In *The Geography of Childhood: Why Children Need Wild Places*, edited by Gary Paul Nabhan and Stephen Trimble. 1994. Boston: Beacon Press.

____. "Going Truant: the Initiation of Young Naturalists." In *The Geography of Childhood: Why Children Need Wild Places*, edited by Gary Paul Nabhan and Stephen Trimble. 1994. Boston: Beacon Press.

Nabhan, Gary Paul and Stephen Trimble. *The Geography of Childhood: Why Children Need Wild Places*. 1994. Boston: Beacon Press.

Nabokov, Peter, ed. *Native American Testimony: A Chronicle of Indian-White Relations from Prophecy to the Present, 1492–1992*. 1978, 1991. New York: Viking.

Nagy, Joseph Falaky. *The Wisdom of the Outlaw: The Boyhood Deeds of Finn in Gaelic Narrative Tradition*. 1985. Berkeley: University of California Press.

Nash, Roderick Frazier. *American Environmentalism: Readings in Conservation History*, 3d ed. 1990. New York: McGraw-Hill.

Naylor, Thomas H., William H. Willimon, and Magdalena R. Naylor. *The Search for Meaning*. 1994. Nashville, Tenn.: Abingdon Press.

Nelson, Peter. *Treehouses: The Art and Craft of Living Out on a Limb*. 1994. Boston: Houghton Mifflin Company.

Nightingale, Benedict. "After 'Chariots of Fire,' He Explores the Legend of 'Tarzan.'" *New York Times*, March 6, 1983.

Nimoy, Leonard. *I am Spock*. 1995. New York: Hyperion.

____. *I am Not Spock*. 1976. New York: Ballantine Nonfiction.

Nollman, Jim. *Spiritual Ecology: A Guide to Reconnecting with Nature*. 1990. New York: Bantam Books.

Nordhoff, Charles and James Norman Hall. *Hurricane*. 1935, 1936, 1963. New York: Bantam Books.

Oates, Whitney Jennings and Charles Theophilus Murphy. *Greek Literature in Translation*. 1944. New York: Longmans, Green and Co.

Oelschlaeger, Max. *The Idea of Wilderness: From Prehistory to the Age of Ecology*. 1991. New Haven, Conn.: Yale University Press.

Olson, Sigurd F. *Reflections from the North Country*. 1976. New York: Alfred A. Knopf.

Orr, David W. *Ecological Literacy: Education and the Transition to a Postmodern World*. 1992. Albany: State University of New York Press.

Orshefsk. "Gone Are the Pals of Yesteryear." *Life*, June 14, 1963, 102.

Ostling, Richard N. "Galileo and Other Faithful Scientists." *Time*, December 28, 1992, 42.

Passmore, John. *Man's Responsibility for Nature*. 1974. London: Gerald Duckworth & Co., Ltd.

Pelikan, Jaroslav. *On Searching the Scriptures — Your Own or Someone Else's: A Reader's Guide to* Sacred Writings *And Methods of Studying Them*. 1992. New York: Quality Paperback Book Club.

Pelikan, Jaroslav, ed. *Confucianism: The Analects of Confucious (Sacred Writings* series). Translated by Arthur Waley. 1992. New York: Quality Paper Back Book Club.

Philippidis, Alex. "Cosmic Controversy: The Big Bang and Genesis 1." *Perspectives on Science & Christian Faith* 47 (1995), 190.

Polkinghorne, John. "Cross-Traffic Between Science and Theology." *Perspectives on Science & Christian Faith* 43 (1991), 144.

Pommersheim, Frank. *Braid of Feathers: American Indian Law and Contemporary Tribal Life*. 1995. Berkeley: University of California Press.

Raglan, Lord. *The Hero: A Study in Tradition, Myth, and Drama*. 1956 (1975 reprint). Westport, Conn.: Greenwood Press.

Rand, Ayn. *The Fountainhead*. 1943, 1971. New York: A Signet Book.

Richmond, Trudie Lamb. "A Native Perspective of History: The Schaghticoke Nation, Resistance and Survival." In *Enduring Traditions: The Native Peoples of New England*, edited by Laurie Weinstein. 1994. Westport, Conn.: Bergin & Garvey.

Rieser, Alison. "Ecological Preservation as a Public Property Right: An Emerging Doctrine in Search of a Theory." *Harvard Environmental Law Review* 15 (1991), 393.

Rockefeller, Steven C. and John C. Elder, eds. *Spirit and Nature: Why the Environment Is a Religious Issue*. 1992. Boston: Beacon Press.

Rockwood, Roy. *Bomba the Jungle Boy; Or the Old Naturalist's Secret*. 1926. New York: Cupples & Leon Company.

Roszak, Theodore. *The Voice of the Earth*. 1992. New York: Simon & Shuster.

Rothenberg, David. "Individual or Community? Two Approaches to Ecophilosophy in Practice." In *Ecological Prospects: Scientific, Religious, and Aesthetic Perspectives*, edited by Christopher Key Chapple. 1994. Albany: State University of New York Press.

Rovin, Jeff. *The Films of Charlton Heston*. 1977. Secaucus, N.J.: The Citadel Press.

Sagan, Carl and Ann Druyan. *Shadows of Forgotten Ancestors: A Search for Who We Are*. 1992. New York: Random House.

Sagan, Dorion and Lynn Margulis. "Gaian Views." In *Ecological Prospects: Scientific, Religious, and Aesthetic Perspectives*, edited by Christopher Key Chapple. 1994. Albany: State University of New York Press.

Sagoff, Mark. "Settling America or the Concept of Place in Environmental Ethics." *Journal Energy, Natural Resources & Environmental Law* 12 (1992), 349.

Sauro, Joan. *Whole Earth Meditation: Ecology for the Spirit.* 1986, 1992. San
    Diego, Calif.: LuraMedia.
Sax, Joseph L. "Liberating the Public Trust Doctrine from Its Historical
    Shackles." *University of California Davis Law Review* 14 (1980), 186,
    188, 192–193.
Sayre, William. "The Rural Economy: Essential to Vermont's Characer and
    Culture." *Vermont's Forests and You* (March 1992), 6.
Schama, Simon. *Landscape and Memory.* 1995. New York: Alfred A. Knopf.
Schickel, Richard. "The Wild Child Noble Savage." *Time,* 1984, 89.
Schneebaum, Tobias. *Wild Man.* 1979. New York: Viking Press.
Schorsch, Ismar. "Learning to Live with Less." In *Spirit and Nature: Why the
    Environment Is a Religious Issue,* edited by Steven C. Rockefeller and
    John C. Elder. 1992. Boston: Beacon Press.
Secrest, Meryle. *Frank Lloyd Wright.* 1992. New York: Alfred A. Knopf.
Sheldrake, Rupert. *The Rebirth of Nature: The Greening of Science and God.*
    1991. New York: Bantam Books.
Shepard, Paul. *Man in the Landscape: A Historic View of the Esthetics of Nature.*
    1967, 1991. College Station, Tex.: Texas A & M University Press.
Silko, Leslie Marmon. *Ceremony.* 1977. New York: The Penquin Group.
Simpson, George Gaylord. *The Meaning of Evolution: A Special Revised and
    Abridged Edition.* 1949, 1951. New York: A Mentor Book.
Siskel, Gene. "'Tarzan': Innocent Bits of Hollywood and Vine." *Chicago
    Tribune,* July 27, 1981, sec. 2, 5.
Slate, Tom. "Edgar Rice Burroughs and the Heroic Epic." *Riverside Quarterly* 3
    (March 1968), 122.
Snoke, David W. "Toward a Unified View of Science and Theology."
    Perspectives on Science & Christian Faith 43 (1991), 172.
Snyder, Gary. *The Practice of the Wild.* 1990. San Francisco: North Point Press.
Stanton, Mark and Dennis Guernsey. "Christians' Ecological Responsibility: A
    Theological Introduction and Challenge." *Perspectives on Science &
    Christian Faith* 45 (1993), 2.
Starhawk, *The Fifth Sacred Thing.* 1993. New York: Bantam Books.
Stegner, Wallace. "The Sense of Place." *Harrowsmith Country Life*
    (September–October 1992), 41.
____. *Where the Bluebird Sings to the Lemonade Springs: Living and Writing in
    the West.* 1992. New York: Random House.
____. "The Meaning of Wilderness for American Civilization." In *American
    Environmentalism: Readings in Conservation History,* 3d ed., edited by
    Roderick Frazier Nash. 1990. New York: McGraw-Hill.
Stevens, Jan S. "The Public Trust: A Sovereign's Ancient Prerogative Becomes
    the People's Environmental Right." *University of California Davis Law
    Review* 14 (1980), 195–196, 199.
Stewart, George R. *Earth Abides.* 1949. Greenwich, Conn.: Fawcett Crest.

Stilgoe, John R. "Defining the Young Nation." *Civilization* (July–August 1995), 64.

Street, Brian V. *The Savage in Literature: Representatives of "Primitive" Society in English Fiction 1858–1920.* 1975. Boston: Routledge & Keegan Paul.

Swimme, Brian and Thomas Berry. *The Universe Story: From the Primordial Flaring Forth to the Ecozoic Era — A Celebration of the Unfolding of the Cosmos.* 1992. New York: HarperSanFrancisco.

Symcox, Geoffrey. "The Wild Man's Return: The Enclosed Vision of Rousseau's *Discourses.*" In *The Wild Man Within: An Image in Western Thought from the Renaissance to Romanticism,* edited by Dudley, Edward and Maximillian E. Novak. 1972. Pittsburgh, Pa.: University of Pittsburgh Press.

Tall, Deborah. "Dwelling: Making Peace with Space and Place." *Orion* (Spring 1995), 16.

Tarlock, A. Dan. "Earth and Other Ethics: The Institutional Issues." *Tennessee Law Review* 56 (1988), 44, 76.

Taylor, Bron. "Earth First!'s Religious Radicalism." In *Ecological Prospects: Scientific, Religious, and Aesthetic Perspectives,* edited by Christopher Key Chapple. 1994. Albany: State University of New York Press.

Thomas, Lewis. *The Fragile Species.* 1992. New York: Charles Scribner's Sons.

Thoreau, Henry David. *A Week on the Concord and Merrimack River — Walden; or, Life in the Woods — The Maine Woods — Cape Cod.* 1985. New York: Literary Classics of the United States, Inc.

Torgovnick, Marianna. *Gone Primitive: Savage Intellects, Modern Lives.* 1990. Chicago: University of Chicago Press.

Trimble, Stephen. "A Land of One's Own." In *The Geography of Childhood: Why Children Need Wild Places,* edited by Gary Paul Nabhan and Stephen Trimble. 1994. Boston: Beacon Press.

Tuan, Yi-Fu. *Passing Strange and Wonderful: Aesthetics, Nature, and Culture.* 1993. Washington, D.C.: Island Press/Shearwater Books.

Tucker, Mary Evelyn. "An Ecological Cosmology: The Confucian Philosophy of Material Force." In *Ecological Prospects: Scientific, Religious, and Aesthetic Perspectives,* edited by Christopher Key Chapple. 1994. Albany: State University of New York Press.

Tucker, William. *Progress and Privilege: America in the Age of Environmentalism.* 1982. Garden City, N.Y.: Anchor Press/Doubleday.

Ulman, Neil. "An Inuit Artist's Self Portrait in Stone." *The Wall Street Journal,* July 14, 1993, A10.

van Dyke, Fred G. "Ecology and the Christian Mind: Christians and the Environment in a New Decade." *Pespectives on Science & Christian Faith* 43 (1991), 174.

Walters, Mark Jerome. "Democracy of Caring." *Natural History* (March 1996), 75.

Waterman, Laura and Guy Waterman. *Wilderness Ethics: Preserving the Spirit of Wildness.* 1993. Woodstock, Vt.: The Countryman Press.

Watson, John H. *The Adventures of the Peerless Peer,* edited by Philip Jose Farmer. 1974. Boulder, Colo.: The Aspen Press.

Weatherford, Jack. *Indian Givers: How the Indians of the Americas Transformed the World.* 1988. New York: Crown Publishers, Inc.

Weinstein, Laurie, Delinda Passas, and Anabela Marques. "The Use of Feathers in Native New England." In *Enduring Traditions: The Native Peoples of New England,* edited by Laurie Weinstein. 1994. Westport, Conn.: Bergin & Garvey.

Weinstein, Laurie, ed. *Enduring Traditions: The Native Peoples of New England.* 1994. Westport, Conn.: Bergin & Garvey.

Weiskel, Timothy C. "The Need for Miracles in the Age of Science." *Harvard Divinity Bulletin* 20 (Summer 1990), unpaginated.

Wenz, Peter S. *Environmental Justice.* 1988. Albany: State University of New York Press.

White, Richard. *"It's Your Misfortune and None of My Own": A New History of the American West.* 1991. Norman: University of Oklahoma Press.

Whitfield, Stephen E. and Gene Roddenbury. *The Making of Star Trek.* 1968. New York: Ballantine.

Williams, Terry Tempest. *An Unspoken Hunger: Stories from the Field.* 1994. New York: Pantheon Books.

Williamson, Ray A. *Living the Sky: the Cosmos of the American Indian.* 1984. Norman: University of Oklahoma Press.

Winn, Marie. "Central Park and Its Bird-Watchers Go Wild." *The Wall Street Journal,* September 8, 1994, A16.

Winokur, L. A. "Pushing Their Luck: Zuni Indians Peddle 'Magical' Charms." *The Wall Street Journal,* April 28, 1993, 1.

Woodley, Richard. *Man from Atlantis.* 1977. New York: Dell Publishing Co.

Wright, Patricia C. "Ecological Disaster in Madagascar and the Prospects for Recovery." In *Ecological Prospects: Scientific, Religious, and Aesthetic Perspectives,* edited by Christopher Key Chapple. 1994. Albany: State University of New York Press.

Wright, Thomas. "Science, God and Man." *Time,* December 28, 1992, 38.

Yarbro, Chelsea Quinn. *Blood Games.* 1979. New York: St. Martin's Press.

____. *Hotel Transylvania.* 1978. New York: St. Martin's Press.

Zeveloff, Samuel I., L. Mikeal Vause, and William H. McVaugh, eds. *Wilderness Tapestry: An Eclectic Approach to Preservation.* 1992. Reno: University of Nevada Press.

Zingg, R. M. "Feral Man and Extreme Cases of Isolation." *American Journal of Psychology* 53 (1940), 487.

Index

New Adventures of Tarzan, The (novelization), 168
Newberry, John Strong, 55
New York City, 25, 117, 200
Niehbuhr, 105
Noble savage, xiii, 33, 44, 47–49, 63–74,
    120, 137, 140, 153, 163, 171, 179, 190;
    as Amerindians, 44, 93–94, 95;
    archetypal, 223–24; Caine as, 190;
    Dionysus as, 48; distaff, 173; in fiction,
    140; in film, xiii, 173, 178; Kioga as,
    165, 166; Rambo as, 178; Roark as,
    180; Rousseau's, 49, 69; Tarzan as,
    147; Terangi as, 166
Nu, 171–72
Nyoka, 173

O'Keeffe, Miles, 155
Olmsted, Frederick Law, 116
Olson, Sigurd, 122
On Deadly Ground, xi, xiii–xiv
Origen, 7
Origin story, 95–96
Oroonoko, 49
Outlaw, 27, 59, 67, 192

Pastoral, xv, 59, 125, 201; pastoral ethos,
    30, 34; pastoral image, 22, 23, 47;
    pastoral myth, 22; pastoral poetry, 45;
    pastoral romance, 56; pastoral setting,
    34; pastoral version of the Simple Life,
    44–46, 59; pastoralism, 25, 116, 117
Peter Blue Cloud, 93
Philo, 7
Philosophy, 30, 43, 55, 58, 63, 87, 114,
    213, 218; eco-philosophy, 228; of First
    Nations, 229; global environmental
    philosophy, 235; of law, 228; natural
    philosophy, 18; in romantic poetry, 57;
    in *Walden,* 55
Physical realm, xvi, 23
Pinchot, Gifford, 23
Place, 34, 35, 47, 55, 63, 109, 117–21,
    123–24, 139, 199–201, 218, 227;
    child's sense of, 118–19; commitment
    to, 233; in popular culture, 131–42
    passim; sacred, 31; sense of, xv, 14, 31,
    114–25; special, 123

Planet of the Apes, 179
Plato, 4, 5–6
Plotinus, 8
Poe, Edgar Allan, 60
Pollution control, xvi, 29, 226
Polynesian peoples, 85
Pommersheim, Frank, 233
Pope John Paul II, 17, 80, 213, 214
Primal culture, xiv, 4, 81–85, 121, 200,
    202; people, xiii, 69, 81; story, 11
Primal hunter-gatherer, 95
Primal image, xiii
Primal memory, 137
Primal mythology, 31
Primitive, (the), 27, 37, 46
Primitivism, 46, 49
Psychology, 31, 215–16; Jung on, 63; in
    romantic poetry, 57
Public trust doctrine, 218, 226

Quest, The, 190
Questor Tapes, The, 192

Rachel Rosen, 185
Rambo, (John), 66, 134, 178–79
Rand, Ayn, 135–36
Rapoport, Amos, 84
Redeemer (archetype), 65
Religion, 4, 6, 16, 18, 36, 44, 47, 63, 80,
    86–88, 204, 205, 210, 213–14, 215.
    See also specific religions
Remington, Frederick, 53
Rene, 69
Rhalina, 184
Rick Deckard (fiction), 185
Roark, (Howard) (fiction), 135–36,
    180–181
Robin Hood, 27, 65
Robinson Crusoe, 69, 183
Roddenberry, Gene, 169–70
Romanticism, 14, 16, 23, 33, 56–57;
    romantic movement, 25, 56–57, 124;
    romantic poet, 74, 138
Roosevelt, Teddy, 23
Roszak, Theodore, 15, 74, 138
Rousseau, 45, 49, 71
Rumi, Jalahad-Din, 80
Rural, 5, 58; rural world, 34